Experiments in International Benchmarking of US Research Fields

Committee on Science, Engineering, and Public Policy

National Academy of Sciences
National Academy of Engineering
Institute of Medicine

NATIONAL ACADEMY PRESS
Washington, D.C.

NATIONAL ACADEMY PRESS • 2101 Constitution Ave., NW • Washington, DC 20418

NOTICE: This volume was produced as part of a project approved by the Governing Board of the National Research Council, whose members are drawn from the councils of the National Academy of Sciences, the National Academy of Engineering, and the Institute of Medicine. It is a result of work done by the Committee on Science, Engineering, and Public Policy (COSEPUP) as augmented, which has authorized its release to the public. This report has been reviewed by a group other than the authors according to procedures approved by COSEPUP and the Report Review Committee.

The Committee on Science, Engineering, and Public Policy (COSEPUP) is a joint committee of NAS, NAE, and IOM. It includes members of the councils of all three bodies.

Financial Support: The development of this report was supported by the Sloan Foundation and the National Research Council.

International Standard Book Number 0-309-06898-3

Internet Access: This report is available on COSEPUP's World Wide Web site at http://www4.nationalacademies.org/pd/cosepup.nsf

Order from: National Academy Press, 2101 Constitution Ave., NW, Washington, DC 20418. *All orders must be prepaid with delivery to a single address. No additional discounts apply. Prices are subject to change without notice. To order by credit card, call* 1-800-624-6242 or 202-334-3313 (in Washington metropolitan area).

Printed in the United States of America.

Committee on Science, Engineering, and Public Policy

COSEPUP Benchmarking Guidance Group

MARYE ANNE FOX *(Chair)*, Chancellor, North Carolina State University
DAVID CHALLONER, Director, Institute for Science and Health Policy, University of Florida
ELLIS COWLING, University Distinguished Professor At-Large, North Carolina State University
GERALD DINNEEN, Retired Vice President, Science and Technology, Honeywell, Inc.
MILDRED S. DRESSELHAUS, Institute Professor of Electrical Engineering and Physics, Massachusetts Institute of Technology
ALEXANDER FLAX, Consultant
RALPH E. GOMORY, President, Alfred P. Sloan Foundation

DEBORAH D. STINE, Study Director
ALAN ANDERSON, Consultant Writer
REBECCA BURKA, Administrative Associate
AUBREY SABALA, Project Assistant
NORMAN GROSSBLATT, Editor

PREFACE

The American people, through their elected representatives, support the nation's research enterprise in the expectation of substantial returns on their investment: a higher standard of living, a healthier society, an environmentally sustainable economy, and a strong national security. Knowing the power of research in addressing national objectives, the nation has committed itself to a broad set of investments to uphold its research capability.

The National Research Council has already prepared studies that describe the effectiveness of research investments in addressing national concerns. Research investments affect the quality of research done. The present study asks how to evaluate research-leadership status. COSEPUP proposed contributing a set of experiments in international benchmarking. International benchmarking compares the quality and impact of research in one country (or region) with world standards of excellence. The use of international benchmarking was also advocated in 1995 by the "Press report," *Allocating Federal Funds for Science and Technology* (see appendix B), for the purpose of providing objective information for the executive branch and Congress. The need for objective evaluations has intensified since the passage in 1993 of the Government Performance and Results Act (GPRA), which requires annual performance reports by all federal agencies, including those which support research. GPRA is discussed in the 1999 COSEPUP report *Evaluating Federal Research Programs: Research and the Government Performance and Results Act*.

Although the use of international benchmarking was not new, it had not been attempted on a scale large enough to contribute to national policy. Accordingly, in 1997, COSEPUP decided to undertake a set of experiments to test the efficacy of international benchmarking. The committee chose three areas of research—mathematics, immunology, and materials science and engineering—that are quite different from one another in size, funding, numbers of subdisciplines, and other qualities. COSEPUP appointed a panel for each field and sought to experiment with providing information for relatively modest commitments of time and money. Once the panels had completed their reports, the committee held a workshop with agency representatives, congressional staff, and oversight bodies to discuss the findings. (See appendix C for a summary of the workshop.)

This report describes the background, methodology, experimental results, and findings of the international benchmarking experiments and concludes that international benchmarking by a panel of experts can be efficient and reasonably objective. International benchmarking might also be a valuable assessment tool for those seeking to implement GPRA.

Maxine Singer
Chair
Committee on Science, Engineering, and Public Policy

ACKNOWLEDGMENTS

This report is the product of many individuals. First, COSEPUP acknowledges those who made presentations at the Workshop on International Benchmarking: Arden Bement, chair, Panel on Materials Science and Engineering; Irving Weissman, chair, Panel on Immunology; Peter Lax, chair, Panel on Mathematics; and the discussants: Richard Russell, House Committee on Science; Don Lewis, director, Division of Mathematical Sciences, National Science Foundation; Patricia Dehmer, associate director, Office of Science, Office of Basic Energy Sciences, Department of Energy; Helen Quill, chief, Basic Immunology Branch, Division of Allergy, Immunology and Transplantation, National Institute of Allergy and Infectious Diseases, National Institutes of Health; and Irwin Feller, director, Institute for Policy Research and Evaluation, and professor of economics, Pennsylvania State University.

The report was reviewed by persons chosen for their diverse perspectives and technical expertise, in accordance with procedures approved by the National Research Council's Report Review Committee. The purposes of this independent review are to provide candid and critical comments that will assist COSEPUP in making its report as sound as possible and to ensure that the report meets institutional standards of objectivity, evidence, and responsiveness to the study charge. The review comments and draft manuscript remain confidential to protect the integrity of the deliberative process. We wish to thank the following for their participation in the review of this report: Arden L. Bement, Jr., Robert M. White, Margaret H. Wright, Paul A. Fleury, Susan Cozzens, Arthur Bienenstock, and the report review coordinator, Anita

Jones, and the report review monitor, John Ahearne. Although those persons provided many constructive comments and suggestions, responsibility for the content of the report rests solely with COSEPUP.

The production of this report was the result of hard work by the committee as a whole and by the extra effort of the Guidance Group, consisting of current and former COSEPUP members Marye Anne Fox (Chair), David Challoner, Ellis Cowling, Gerald Dinneen, Mildred S. Dresselhaus, Alexander Flax, and Ralph E. Gomory.

The project was aided by the invaluable help of COSEPUP professional staff: Deborah D. Stine, associate director of COSEPUP and study director; Rebecca Burka, administrative associate; Aubrey Sabala, project assistant; Alan Anderson, consultant writer; and Norman Grossblatt, editor. The committee also thanks Richard Bissell, executive director of the Policy Division and director of COSEPUP, who oversaw the committee's activities.

CONTENTS

EXECUTIVE SUMMARY

In its report *Science, Technology, and the Federal Government: National Goals for a New Era*,[1] the Committee on Science, Engineering, and Public Policy (COSEPUP) of The National Academies recommended that the status of research fields might be evaluated by the use of international benchmarking. COSEPUP was referring to a technique that is designed to compare one country's or region's leadership status in research with world standards of excellence.

To assess the feasibility and utility of international benchmarking, COSEPUP proposed a set of experiments in three fields: mathematics, immunology, and materials science and engineering. The results of the experiments suggest that research leadership status by field can be assessed in a timely fashion at reasonable cost. International benchmarking has potential utility for policy-makers and federal agencies in determining the leadership status of the United States in particular fields, subfields, and possibly sub-subfields. It might also help federal agencies to comply with the Government Performance and Results Act by evaluating the quality of their own performance.

Committee members note especially the following points:

- Panel leaders began the process as skeptics but came to believe that benchmarking can be reasonably objective and timely.
- Two separate benchmarking experiments in mathematics—one at the National Academies, the other at the National Science Founda-

[1] Known as the "Goals" report.

tion—despite dissimilar panels and mandates produced similar results, lending credence to the technique.

During the experiments, committee members were able to identify several particular strengths of international benchmarking:

- Panels were able to identify institutional and human-resource factors crucial to maintaining leadership status in a field that is unlikely to have been identified by other methods.
- Benchmarking allows a panel to determine the best measures for a particular field while providing corroboration through the use of different methods, as opposed to the "one-size-fits-all" approach of some common evaluation methods.
- Benchmarking can produce a timely but broadly accurate "snapshot" of a field.

The experiments were sufficiently thorough to provide guidelines for future experiments, including the following:

- Because of the panels' use of expert judgment, the choice of panelists is a key to the credibility of the results. A tendency toward national biases can be mitigated by ensuring diverse geographic membership of panels; the same is true of the groups that select the panel members. In particular, it is critical to include non-US participants in the selection of panelists and as panel members because they provide perspective and objectivity.
- Because major fields of research change slowly, benchmarking can probably detect important changes in quality, relevance, and leadership in fields when conducted at intervals of 3-5 years. It is unlikely that changes can be detected by annual benchmarking.
- The choice of research fields to be evaluated is both challenging and critical. A "field" might best be considered the array of related domains between which investigators can move without leaving their primary area of expertise.
- Benchmarking produces information that administrators, policy-makers, and funding agencies find useful as they make decisions as to what activities a federal research program should undertake and respond to demands for accountability, such as the Government Performance and Results Act.
- If federal agencies use benchmarking, the wide variation in agency missions dictates that each agency tailor the technique to its own needs.
- Use of indicators that provide information on degree of uncertainty and reliability of benchmarking results might enhance the presentation of panel assessments of leadership status.

The present experiments should be regarded as an encouraging first step toward the design of an efficient and reliable evaluation tool. An accumulation of benchmarking exercises will lead to more-effective methods, better understanding of the factors that promote research excellence, and better decision-making by funders of science and technology.

1

INTRODUCTION

In 1992, the Committee on Science, Engineering, and Public Policy (COSEPUP) of the National Academies examined the federal government's rationale for funding research in science and technology. There is little doubt that science and technology are powerful tools in moving the nation toward social and economic goals,[2] and yet it has proved difficult to devise a consistently fair and effective mechanism for allocating federal funds for research.

National policy-makers desire better mechanisms for several reasons. From the funder's point of view, it is natural to ask what the outcomes of a particular program or project are likely to be and when they might be expected. But it is seldom possible to predict the outcomes of basic research or to know which fields of research will ultimately contribute to important new ideas or technologies.

At the same time, there is little debate about the necessity for sustained agendas of both basic and applied research.[3] Retrospective studies consistently describe the power of research to produce breakthroughs years and even decades after the work has been performed.[4] Therefore, the committee has no doubt about the need for sustained federal funding of basic research in every major field.

---

[2]COSEPUP, *Capitalizing on Investments in Science and Technology,* 1999.

[3]COSEPUP, *Evaluating Federal Research Programs: Research and the Government Performance and Results Act,* 1999.

[4]Donald Stokes, *Pasteur's Quadrant: Basic Science and Technological Innovation.* Washington, DC: Brookings Institution Press, 1997.

When the committee issued its "goals report" (*Science, Technology, and the Federal Government: National Goals for a New Era*),[5] it recommended that the federal government continue vigorous funding of basic research and that it seek to support this research across the entire spectrum of scientific and technological investigation. That advice was justified by the fact that leadership in science had become one of the defining characteristics of the United States and other great nations.

Specifically, COSEPUP suggested two goals. First:

The United States should be among the world leaders in all major areas of science.

"Major areas" refers to broad disciplines of science (such as biology, physics, mathematics, chemistry, earth science, and astronomy) and to their major subdisciplines (such as the neurosciences, condensed-matter physics, and seismology). "Among the world leaders" means that the United States should have capabilities and infrastructures of support that are not substantially exceeded elsewhere. The primary rationale for the recommendation is that working at a world standard of excellence in all fields allows this nation to apply and extend scientific advances quickly no matter when or where in the world they occur. The value of being among the leaders was dramatized when, for example, the phenomenon of high-temperature superconductivity was demonstrated in an international laboratory in Switzerland. US researchers, although they did not participate in that breakthrough, were able to replicate the results in a matter of weeks because they were working at the frontiers of solid-state physics (the general field in which the breakthrough occurred) and they had the ability to move quickly. In addition, because of the degree of interconnection between fields, there is a concern that if one were neglected (placing the United States behind the world leaders), others might be slowed as a result. For example, much of the progress in life sciences research is made possible by the availability of instruments designed by scientists and engineers in physics and chemistry.

Second:

The United States should maintain clear leadership in some major areas of science.

Such areas would include those which are required to meet national objectives, which capture the imagination of society, or which have multiplicative effects in other important fields. For example, the United

[5]National Academy Press, Washington, DC, 1993. See excerpt in Appendix B-1.

States desires to maintain clear leadership in molecular genetics because of its central importance to human health and to the biotechnology industry.

Both those goals were reiterated in another Academies report, *Allocating Federal Funds for Science and Technology*,[6] developed by a committee chaired by former National Academy of Sciences President Frank Press which stated that "to continue as a world leader, the United States should strive for clear leadership in the most promising areas of science and technology and those deemed most important to our national goals. In other major fields, the United States should perform on a par with other nations so that it is 'poised to pounce' if future discoveries increase the importance of one of these fields."

In the goals report, COSEPUP considered the question of how to measure leadership. How can the federal government gauge the overall health of scientific research—as a whole and in its parts—and determine whether national funding adequately supports national research objectives? The committee wrote that it is feasible to monitor US performance with field-by-field peer assessments and that this might be done through

> ***the establishment of independent panels consisting of researchers who work in a field, individuals who work in closely related fields, and research "users" who follow the field closely. Some of these individuals should be outstanding foreign scientists in the field being examined.***

This technique of comparative international assessments, or "international benchmarking", had been discussed in theory and applied in specific cases. But COSEPUP made the decision to undertake a set of realistic experiments that would test the utility of the technology in evaluating entire research fields.

The committee acknowledged that quantitative indicators commonly used to assess research programs—for example, dollars spent, papers cited, and numbers of scientists supported—are useful information but noted that by themselves they are inadequate indicators of leadership, both because quantitative information is often difficult to obtain or compare across national borders and because it often illuminates only a portion of the research process. For example, a paper that describes truly innovative research may receive few citations if no one else is doing comparable work. Similarly, the pervasive value of mathematics to the other sciences is not accurately measured by the small number of dollars flowing into mathematical research.

[6] National Academy Press, Washington, D.C., 1995. See excerpt in Appendix B-2.

The committee decided that the expert judgment of panel members afforded the most effective means for assessing research. The basis for this decision is that those immersed in a particular field are best qualified to appraise the quality of its researchers, identify the most promising advances, and project the status of the field into the future. Such panelists are also well positioned to pinpoint locations where the most promising ideas are emerging, describe where the best new scientific talent chooses to work, and judge the comparative quality of research facilities and human resources. The participation of representatives of other countries, people in other fields of research, and groups that use the results of research helps to provide perspective and objectivity to the exercise and to keep the panels' judgments from being self-serving.

In addition, passage of the Government Performance and Results Act (GPRA) in 1993 emphasized the need for a method to assess the results of scientific and technological research investments. Although COSEPUP began its deliberations about benchmarking before GPRA became law, it recognized while developing its report that benchmarking might provide input that is useful to some federal agencies as they evaluate their own research programs. The committee's GPRA report[7] concluded that the most effective means of evaluating federal research is expert review. COSEPUP defined expert review in terms of three elements—quality, relevance, and leadership—and proposed that leadership be evaluated via international benchmarking. A benchmarking panel would be asked, "Is the research being performed at the forefront of scientific and technological knowledge?" The panel would then consider the mission objectives of the particular agency in the context of an assessment of the nation's scientific leadership status.

In the language of COSEPUP's GPRA report:

> For agencies whose missions include a specific responsibility for basic research—such as the National Science Foundation in broad fields of science and engineering, the National Institutes of Health in fields related to health, or the Department of Energy in high-energy physics—world leadership in a field can itself be an agency goal. That is equally true for mission agencies, such as the Department of Defense (DoD), but in more focused ways. For example, DoD can take as a goal world leadership in basic materials research relevant to its mission. Once such a goal is established, the usual measures of quality and leadership should be applied.

In this report, COSEPUP evaluates the feasibility and utility of the benchmarking technique. As explained in greater depth in chapter 2, it

[7] *Evaluating Federal Research Programs: Research and the Government Performance and Results Act,* National Academy Press, Washington, DC, 1999. Benchmarking is discussed briefly in the report as it relates to GPRA.

does so by internationally benchmarking three fields: mathematics, immunology, and materials science and engineering. The results of these experiments are summarized in chapter 3. The committee held a workshop to obtain input from agencies, policy-makers, disciplinary societies, and others on the feasibility and utility of benchmarking; the workshop provided material for chapter 4, in which COSEPUP develops its findings, and chapter 5, where those findings are discussed. Conclusions are listed in chapter 6.

2

METHODOLOGY

Once COSEPUP decided on the use of expert panels for benchmarking, it found few models on which to base a method. Traditional peer-review panels provided a sound precedent, although most peer reviewers focus only on the quality of a program or project. They might or might not be asked to assess quality relative to world standards, to identify the key determinants of research performance, or to assess the future performance of research programs. The latter exercises, however, are integral to international benchmarking and require experts who have a broad understanding of a field as a whole and knowledge of the researchers who are most influential in that field.

2.1 General Features of Methodology

COSEPUP decided that the benchmarking of each field should be guided by an oversight group that included people with broad backgrounds. The oversight group, which included US members of relevant National Research Council commissions and boards, was asked to define a field's subfields (which could later be modified by the panel) and to select a benchmarking panel of the highest quality. In addition, because each panel was part of an overall study, it was important to strive for a consistent approach. COSEPUP offered the following guidelines to each panel before it began its work:

- Panels should develop findings and conclusions regarding the US leadership status in a research field, but not recommendations. In particular, the panel members should avoid statements that might be construed as recommendations to increase the funding for their field.

- Each panel, in developing its report, should seek a consensus based mainly on the informed judgments of panel members.
- Panel members should focus on the accomplishments of researchers in the field, not on funding levels, as indicators of leadership and they should consider the development of human resources as a component of leadership; for example, if all the interesting research in a country is being done by senior researchers, that country might lack sufficient young researchers to develop accomplishments in the future.

2.2 Specific Charge to the Panels

In particular, COSEPUP charged each panel to answer three questions:

1. *What is the position of US research in the field relative to that in other regions or countries?* Is it at the forefront? The leader? Behind the leaders? Where is the most exciting research occurring internationally? Given that the United States should be at the forefront of research, where does US research stand relative to the forefront?
2. *On the basis of current trends in the United States and worldwide, what will be our relative position in the near and longer-term future?* Given current trends, will US research in the field remain at the forefront? Take the lead? Fall behind the leaders?
3. *What are the key factors influencing relative US performance in the field?* Why is the critical research occurring either in or outside the United States? Is the equipment, infrastructure, or supply of young people superior or inferior?

Each panel adapted this charge to the characteristics of its particular field. Complete descriptions of methodology can be found in the reports of the three panels, which are in the attachments.

2.3 Selection of Panel Members

To achieve balance and diversity, COSEPUP asked the oversight committees to include panel members from the following general groups: US experts in the research field being assessed, experts in related fields of research, non-US experts in the field and related fields, and "users" of research results.

The concept of "users" is meant to embrace those who can judge both the quality of research and its relevance to further research, to industrial applications, and to other societal objectives, including the advancement of knowledge. These users might be found in academe, government, industry, or other sectors. Users of mathematical research, for example, might include an academic chemist, an industrial engineer, a foreign mathematician, an economist, and a representative of a professional society. COSEPUP also decided to seek an expert in policy analysis for each panel. With this diverse composition, each panel was equipped to judge the quality, relevance, and leadership status of the research field under study.

In addition, the oversight committees tried to create panels that represented geographic and professional diversity. They sought multinational viewpoints to enhance objectivity and reduce national bias. For example, the immunology panel selected "up-and-coming leaders as well as established ones, investigators from all over the world, and leaders of all sub-subfields, both basic and clinical" to "incorporate the opinions of a variety of respected members of the immunology community." Membership from industrial firms was deemed essential to provide perspective on the work of leading industrial researchers and on the ultimate utility of research to developmental and manufacturing processes on which firms depend. Inclusion of younger and senior researchers were seen as essential to provide innovative thinking and fresh perspective.

2.4 Selection of Research Fields

For these experiments, the committee deliberately chose difficult subjects—fields that are diverse in scope and subject matter. Mathematics is the closest to being a traditional discipline, but even mathematics is broad in the sense that it is the language and tool of most of the sciences.

Immunology is not a disciplinary field in the traditional sense. Although immunologists work in virtually every department and division of the life sciences, few universities have departments of immunology. Immunology embraces many disciplines—including biochemistry, genetics, and microbiology—and its findings translate into diverse clinical subjects, such as rheumatology, surgery, endocrinology, neurology, and allergy.

Materials science and engineering spans multiple disciplines concerned with the structure, properties, processing, and performance of materials. Nearly all fields of science and engineering are involved in some way with materials, and many ideas in materials science and engineering emerge from disciplines as diverse as solid-state physics, chemistry, electronics, biology, and mechanics.

The process of selecting fields reminded the committee how difficult it is to divide research activities into discrete categories but this is a key to effective benchmarking. There is probably an optimal size and complexity of fields to be benchmarked, but these early experiments could give only rough indications of those measures. Two of the fields, immunology and materials, were initially judged by their panel chairs to be too large for truly rigorous treatment (they later changed their minds).

As scientific research becomes more interdisciplinary and complex, scientists and engineers are challenged to describe the limits of their own intellectual activity. It seems likely that the definitions required by benchmarking exercises can help to illuminate criteria for defining fields and subfields. In the case of superconductivity mentioned above, American scientists who were already working in adjacent subspecialties were able to move quickly into superconductivity research, and this could offer a definition of the subfields within a field along functional lines; that is, a field might be defined as the array of related domains among which investigators can move without leaving the realm of their expertise.

2.5 Evaluation of Panel Results

COSEPUP evaluated the quality of panel results independently and via comments from the oversight groups. In addition, the feasibility and utility of benchmarking were assessed during meetings with disciplinary societies and at a full-day workshop attended by representatives of federal agencies, universities, Congress, and the executive branch. The summary of that workshop in appendix C provided valuable information for COSEPUP members.

Of particular importance were the contributions of reviewers, whose array of expertise was comparable with that of the panel. The reviewers, chosen to represent diverse industrial and academic backgrounds, provided invaluable commentary and criticism for use by the panels and a means of validating the panels' findings.

Among the topics proposed for further discussion were the use of benchmarking during the budgetary process and whether benchmarking would be useful in helping to set national science policy. In addition, federal-agency representatives were asked about their own procedures for evaluating research and whether benchmarking might have a role to play in those procedures. Finally, the relevance of benchmarking to GPRA was discussed.

3

RESULTS OF THE BENCHMARKING EXPERIMENTS

COSEPUP charged each oversight committee and its panel with the general task of benchmarking its own field. The activities of the groups varied somewhat by field, but the overall approaches chosen by the groups were similar. These approaches are summarized in this section.

3.1 How Panels Were Selected

As expected, the members of each oversight group sought out panel members who had broad understanding of their field and extensive connections with the international research community, including members of the academic, industrial, and government sectors.

The attempt to use discrete categories was complicated by overlap. For example, any person is likely to be both a researcher and a user of research. Similarly, some of the "US" academic researchers were born outside the United States or conducted collaborative research with non-US researchers. Still, about half the panel members selected were US academic researchers in the field. The remaining half were US researchers in related fields, "users" of research (as defined in section 2.3), and international (non-US) researchers.

Of the 12 members of the mathematics panel, three were non-US researchers, including a current and a former president of the respective national academies of sciences in their countries. Two were US researchers employed by industry, and one was a US Nobel Prize-winning chemist who uses mathematics in his research. The remaining six members were US academic researchers in mathematics.

Similarly, the 13 members of the materials science and engineering panel included three non-US researchers, two US researchers-administrators in industry—all in materials science and engineering—and one US researcher in a related field. The remainder were US academic researchers in materials science and engineering.

The 14 members of the immunology panel included three non-US researchers, two US researchers-administrators in industry, and one US policy analyst, in addition to US academic researchers.

The oversight groups, charged with nominating panel members and overseeing the benchmarking experiments, were also diverse in membership, although they did not contain foreign members. Each comprised about six individuals. Two members of each oversight group were members of COSEPUP, and the remainder were representatives of related NRC commissions and boards. For example, the oversight group of the materials science and engineering panel included two COSEPUP members and members of the Commission on Physical Sciences, Mathematics, and Applications (CPSMA), the Commission on Engineering and Technical Systems (CETS), CETS's National Materials Advisory Board (NMAB), and CPSMA's Board on Physics and Astronomy (BPA); among them were four industry researchers and three academic researchers, all from the United States.

3.2 How Panels Assessed Their Fields

Each panel used a variety of methods to assess its field; the methods depended on the disciplines within the field. The methods used included

- The "virtual congress".
- Citation analysis.
- Journal-publication analysis.
- Quantitative data analysis (for example, numbers of graduate students, degrees, and employment status).
- Prize analysis.
- International-congress speakers.

Each method is described in more detail below. The assessments were to be current—that is, the status of US research in the field today, not in the past (most analysis relied on information collected within the past 5-10 years). Some information, such as that provided by the virtual congress, were current.

A challenging aspect of the benchmarking exercise was the great size of the US research enterprise. In the three benchmarked fields, US researchers were found to perform a dominant portion of the world's research, as measured by numbers of publications. Because of this size dominance, the panels often drew comparisons of leadership status in relation to regions or to the rest of the world combined rather than to individual countries.

Another challenge for the panels was to determine which subfields to analyze. Panel members used their own judgment and existing documents. For example, the mathematics panel used the meeting sections of the International Congress of Mathematics and the International Congress of Industrial and Applied Mathematics as starting points. The materials science and engineering panel started with subfields identified in a recent report by the National Science and Technology Council that discussed materials science and engineering across the federal government. The immunology panel faced a greater difficulty: the field is an amalgam of subfields in larger disciplines, and no subfield classification systems existed, so the panel chose to develop its own.

A method of defining subfields that reflected the original purpose of the COSEPUP leadership recommendations was to define subfields as areas within which researchers can move and still work within their realm of expertise. That connects directly with the goal of leadership because it enables a country that is strong in a given subfield to respond to and enter a "hot" new area; US researchers were able to do this, for example, in many areas of materials science and engineering.

3.2.1 The Virtual Congress

A technique used by all the panels was to assess leadership by creating an imaginary international "virtual congress". Each panel asked leading experts in the field to identify the "best of the best" researchers for particular subfields, anywhere in the world. That was possible because most subfields were relatively small and their outstanding leaders were well known to active researchers regardless of location.

For example, the immunology panel divided immunology into four major subfields: cellular immunology, molecular immunology, immunogenetics, and clinical aspects of immunology. It divided each of those subfields into four to10 sub-subfields. The panel then identified five to 15 current respected leaders in each sub-subfield and polled each leader in person, by telephone, or by mail, taking into account the number of US and non-US researchers polled. The pollees were asked to imagine themselves as organizers of a session on their particular sub-subfield and to furnish a list of 5-20 current desired speakers. (See table 2.1 in the immunology benchmarking report in attachment III.)

The materials science and engineering panel varied the method to accommodate its larger field. For each of nine subfields, panel members asked colleagues to identify five or six current hot topics and eight to 10 of the best people in the world. The information was used to construct tables that characterized the relative position of the United States in each of the subfields now and projected into the future (see appendix B of the materials benchmarking report in attachment II of the present report).

The first half of each table ranked the current US position relative to the world materials community for each subfield. For scoring purposes, 1 represented the "forefront", 3 represented "among world leaders", and 5 represented "behind world leaders." The second half of each table was an assessment of the likely future position of the United States relative to the world materials community. Here, 1 represented "gaining or extending", 3 represented "maintaining", and 5 represented "losing".

3.2.2 *Citation Analysis*

Citation analysis is a technique often proposed to evaluate the international standing of a country's research in a field. Each panel used an existing British analysis to evaluate US research quality.[8] The analysis included both the numbers of citations and the "relative citation impact," which compares a country's citation rate (the number of citations per year) for a particular field to the worldwide citation rate for that field. A relative citation impact greater than 1 shows that the country's rate for the field is higher than the world's and is viewed by some as a reliable indicator of the quality of the average paper. This latter measure takes into account the size of the US enterprise relative to that in other countries.

The immunology panel also commissioned a "high-impact" immunology database from the Institute for Scientific Information (ISI). (See *http://www.isnet.com/products/rsg/impact.html* for more information on high-impact papers.) The ISI database was scanned for the years 1990-1997. For each year, the 200 most-cited papers in journals relevant to immunology were selected. The 174 authors who had more than five papers on the list were ranked according to the average number of citations per paper (70.2-38.5 citations per paper). (See section 2.1.2 and 2.2.2 of attachment III.) That technique was not used by the other panels, because the degree to which citations are a critical component of leadership status is less in their fields, particularly in materials science and engineering.

3.2.3 *Journal Publication Analysis*

The immunology panel developed a new method called "journal publication analysis." The panel identified four leading general journals (*Nature*, *Science*, *Cell*, and *Blood*) and one of the top journals that focused specifically on immunology (*Immunity*). Panel members

[8] See, for example, *The Quality of the UK Science Base*, United Kingdom, Office of Science and Technology, Department of Trade and Industry, March 1997.

scanned the tables of contents of each of the journals. In the general journals, they identified immunology papers and the laboratory nationality of the principal investigator; and in all the journals, they identified subfields. That allowed a quantitative comparison between publications by US-based investigators and publications by investigators based elsewhere. (See tables 2.3-2.8 and sections 2.1.3 and 2.2.3 of attachment III.)

3.2.4 Quantitative Data Analysis

All the panels encountered difficulty in locating suitable unbiased information by which to compare the major features of the scientific enterprise (such as education, funding and government agencies) in different countries. For example, undergraduate and PhD degrees in the United States are not directly comparable with all similarly labeled degrees in other industrialized countries.

In some fields, notably mathematics, quantitative information that could be used for international comparisons was available from the National Science Foundation (NSF). However, in other countries, comparable information is rare and limited in focus. Panelists found that information on some issues (such as trends in the numbers of graduate students and funding for particular fields) was available only for the United States. (See section 5.5 and appendix B of the mathematics benchmarking report in attachment I of the present report.)

3.2.5 Prize Analysis

Each of the panels analyzed the key prizes given in its field. In mathematics, the key international prizes are the Fields medal and the Wolf prize. In materials science and engineering and in immunology, there are a variety of international prizes. (See section 3.1.1 of attachment II and table 3.1 of attachment III.) In each case, the numbers of non-US and US recipients of these medals were analyzed relative to the location where they now conduct their research. For example, a researcher who originally conducted research in Australia but is now at a US university would be counted as a US researcher; the analysis reflects the current, not past, residence status of a researcher in a country.

3.2.6 International Congress Speakers

Another condition that can be quantified is representation among the plenary speakers at international congresses (not virtual congresses). Although conference organizers strive for geographic balance by inviting speakers from different countries, speakership is an interesting indicator if the US representation at congresses is the same as, smaller, or larger than that in publications generated in a field. (See section 3.1.1 of attachment I.)

3.3 How Panels Characterized US Research

Each panel concluded that the United States was at least among the world leaders in its field. However, each panel also identified subfields in which the United States lagged the world leaders. Each panel identified key infrastructure concerns. The mathematics panel found that the United States was the world leader in the field but that there were "storm clouds on the horizon" because of its heavy reliance on foreign mathematicians who had recently immigrated to the United States. If that influx wanes, the US leadership might be in jeopardy, given the decline in the number of American students applying to graduate school in mathematics. (Although the other panels did not explicitly make that point, more than half the graduate students in science and engineering are foreign-born, so the national research enterprise depends heavily on researchers from other countries.)

The materials science and engineering panel found that the United States was at least among the world leaders in all subfields of materials science and engineering and the leader in some subfields, although neither the United States nor any other country was the world leader in the field as a whole. A general area of US weakness for most subfields was in materials synthesis and processing. Of particular concern in this field was the lack of adequate funding to modernize major research facilities in the United States, many of which are much older than those in other countries, and to build the new facilities needed to maintain research leadership.

The immunology panel found that the United States was the world leader in immunology but that, although US dominance was evident in the major subfields, the United States was only among the world leaders in some parts of subfields. One general concern was the increasing cost of maintaining the mouse facilities that provide a key portion of research infrastructure. Another was the increasing difficulty in locating sufficient numbers of patients for clinical trials.

3.4 Factors Influencing US Performance

Each panel identified the key factors influencing US performance in its field and assessed the factors primarily on the basis of its members' judgment. No quantitative measures were identified that were sufficiently comprehensive to account for performance.

The three panels differed somewhat in their lists of factors considered most important, as might be expected among three widely different fields. However, the three lists overlapped in the following ways:

- Human resources and graduate education—mathematics, materials, and immunology.

- Funding—mathematics, materials, and immunology.
- Innovation process and industry—materials and immunology.
- Infrastructure—materials and immunology.

COSEPUP finds that degree of overlap to be significant, partly because other recent National Academies reports[9] on science and engineering have identified the importance of these same factors.

3.5 Future Relative Position of US Research

With respect to the likely future position of the United States in research relative to other countries, a more diverse set of factors was identified, largely by qualitative techniques:

- Intellectual quality of researchers and ability to attract talented researchers-mathematics, materials, and immunology.
- Ability to strengthen interdisciplinary research-mathematics and materials.
- Maintenance of strong, research-based graduate education-mathematics, immunology, and materials.
- Maintenance of a strong technological infrastructure-materials.
- Cooperation among government, industrial, and academic sectors-materials.
- Increased competition from Europe and other countries-immunology and materials.
- Effect of the shift toward health-maintenance organizations on clinical research-immunology.
- Adequate funding and other resources-mathematics, materials, and immunology.

[9] Examples of these reports include Science, Technology, and the Federal Government: National Goals for a New Era, Evaluating Federal Research Programs: Research and the Government Performance and Results Act, and Allocating Federal Funds for Science and Technology.

4

FINDINGS

In general, COSEPUP judged that the benchmarking experiments were successful. The committee agreed that the technique of benchmarking was able to provide responses on each of the three topics listed in section 2.2 (the relative position of US research today, the relative position of US research in the future, and the key factors influencing relative US performance).

At the outset, some committee members and policy leaders doubted that complex fields of modern science could be assessed to any degree of accuracy without relatively large investments of money and time. However, all three experiments were concluded in a year or less for relatively modest investments.

Panelists found that it is possible to take a "snapshot" of a field (to conduct a leadership survey) by means of a virtual congress in a matter of weeks. This implies that the method shows promise and that further experiments to optimize scale and technique are justified.

COSEPUP found good correlation among the findings produced by various indicators. For example, the qualitative judgments elicited by the virtual congress were similar to the results of quantitative indicators, such as publications cited and papers delivered at international congresses.

In this section, COSEPUP reports its findings with respect to the objectives of the experiments, the composition of panels, the methods used, and the likely utility of benchmarking for federal agencies and policy-makers in the executive branch and Congress. These findings are based on the committee's benchmarking experiments, the comments of reviewers, and the observations of workshop participants (see appendix C for a summary of the workshop).

4.1 Findings About Objectives

When the benchmarking studies were proposed, it was not clear whether a panel of experts in a particular field could analyze that field in an objective fashion. By "objective", the committee means "a reasonably balanced view of US research compared with that of the rest of the world." In spite of that concern, COSEPUP has concluded that each panel has been able to produce a reasonably objective report. An important goal of the benchmarking program was to conduct the studies at modest expense within a short period. Given the large number of fields in science and engineering, benchmarking would not be practical if it were expensive or time-consuming. The three experiments were completed in 6-8 months at a cost of $50,000 each. (A key factor in the low cost is that all panelists agreed to be pro bono contributors.)

4.2 Findings About Results

The studies succeeded in identifying key factors that influenced the status of fields. For example, in mathematics, human resources—particularly the reliance on foreign talent—was identified as a key issue. The panel noted that current US leadership depends substantially on temporary waves of immigrants from Europe (notably from the former Soviet Union) and Asia that cannot be counted on to continue. In addition, the panel warned that the quality of US research could be affected in the future by the observed falloff in numbers of American students pursuing graduate-level mathematics.

Using information from the Department of Energy and the National Institute of Standards and Technology, the materials science and engineering panel was able to identify facilities and infrastructure as the keys to research leadership. In the United States, some materials research facilities, many of which were built in the 1960s, are deteriorating. In Europe and Japan, facilities tend to be more modern.

In immunology, the three tools for evaluating US research (reputation survey, citation analysis, and journal-publication analysis) were distinct and had different strengths and flaws, but they led to basically the same conclusion: the names of US researchers appeared between 2-3 times as often as the names of non-US researchers. The immunology panel was able to identify an important concern that arises from shifts in the US health-care system. The new emphasis on managed care means that fewer patients are available to academic institutions for clinical trials in immunology. The US health-care system differs from that of many European countries, where the centralized medical system provides an abundance of such patients.

4.3 Findings About the Methodology of Benchmarking

COSEPUP found that each panel developed its conclusions by using a similar set of methods, with some variation to match field-specific differences. All panels used the leadership survey, or virtual congress, in which panel members asked colleagues in the United States and abroad whom they would choose to speak at international conferences in particular subfields. Panel members found this method to be the most efficient and credible way to evaluate their fields. People chosen to name the keynote speakers in the virtual congresses tend to be those who have generated the central ideas in their fields and so are in the best position to understand the relative contributions of different scientists and countries. These experts are also in a position to interpret the validity of the quantitative measurements used for benchmarking.

Although the virtual congress does not constitute a systematic assessment and is somewhat subjective, it is the same approach used by leaders of a field to organize real conferences featuring the "best of the best". In conducting this analysis, panel members felt secure in relying not only on their own judgment but also on the judgment of their colleagues and on the collective wisdom of the field. Another advantage of the virtual congress is its swiftness: results are available soon after the experts are polled. One way to test the soundness of this method would be to compare the outcomes of two surveys done in two countries; the outcomes should be the same or very similar. The exercise would also test the likelihood that a country holding the survey is biased in favor of its own researchers.

An interesting overall comparison is available because two independent tests were done in mathematics—one by the COSEPUP mathematics benchmarking panel and another by an NSF panel. The panels were different in composition and in their charges (one was instructed to produce recommendations, and the other was asked not to), but they developed similar sets of conclusions regarding the overall stature of US mathematics and each of its subfields.

Journal analysis proved useful, but its value was mitigated by the amount of time required to analyze information and the need for analysts who were knowledgeable about the field on a worldwide basis. An additional option, not undertaken because of cost constraints, would have been a search of citations in US patent literature for scientific background and prior art.

The use of quantitative measures generally is hampered by the scant availability of comparable international information. The international information that is available is field-dependent and not very timely, and there are variations in the delineation of fields that make comparisons difficult.

4.4 Findings About the Membership of Panels

COSEPUP found that the use of panels is effective when panel members and the experts whom they consult are the most respected innovators in their fields. As leaders, they are in a unique position to understand current developments and trends. The committee also found that the geographic diversity and professional diversity of panel membership are essential to ensure a fair and comprehensive assessment. Over the course of the studies, panels came to agree that no more than half their members should be US academic researchers. The immunology panel, for example, found in its initial response a clear bias related to the laboratory location of the pollees: US-based investigators routinely named a higher percentage of Americans than did non-US-based investigators. The nationality of the poller also appeared to have an influence: the three non-US pollers often obtained a virtual-congress list with a higher proportion of non-US speakers. The panel decided in its second iteration that it needed to increase foreign representation to ensure objectivity; on doing so, it obtained results that agree more closely with those of citation analysis and journal-publication analysis and with the judgment of the panelists.

The committee concluded that at least one-third of panel members should be non-US researchers. An additional one-third should be a combination of researchers in industry and in related fields who use the results of research. In the experience of the panels, that mix of perspectives, including especially the representatives of research-intensive industries (such as biotechnology, telecommunications, and aerospace), was essential for understanding not only the scholarly and technical achievements of researchers, but also the broader importance of those achievements to social and economic objectives.

4.5 Findings About the Utility of Benchmarking

On the basis of presentations made at the workshop by congressional and agency staff and feedback from disciplinary-society members who were briefed by panelists, COSEPUP found that benchmarking is potentially useful to the research communities in selected fields and in fields related to them, and to the government sponsors of research in selected and related fields. The panel reports were able to identify weaknesses in particular subfields and sub-subfields and to point out issues that need to be addressed in making policy.

The committee also suggests that benchmarking might be useful in efforts to comply with GPRA. Representatives of federal agencies that support research were asked during the benchmarking workshop about the utility of the technique. Although the terminology and, to some extent, the concept were new to some, there were some indications that benchmarking was likely to be useful in evaluating agency research programs and in providing information that would help them to comply with GPRA.

5

DISCUSSION

COSEPUP identified several key strengths, weaknesses, and other factors that influence the success of international benchmarking in evaluating research.

5.1 Some Strengths of Benchmarking

International benchmarking produces information that is, in the opinion of COSEPUP, valuable and relevant to researchers, administrators, and policy-makers. Three examples are the heavy reliance of mathematics leadership on foreign talent, the 20- to 30-year age difference between materials equipment in the US and that in several other countries, and the influence of managed care on clinical research. Although most information did not contradict prevailing views, it is unlikely that the results of each report could have been achieved as efficiently by any other technique, given the paucity of the data and information required for a traditional quantitative approach.

International benchmarking is rapid and inexpensive compared with procedures that rely entirely on the assembly of quantitative information. The use of qualitative judgments also has merit. In the words of one panel chair, the panels were able to get "80% of the value in 20% of the time" for a far lower cost.

5.2 Some Weaknesses of Benchmarking

In retrospect, the experiments revealed several methodologic weaknesses that can be addressed in future benchmarking activities. For example, non-US members should be included in the oversight group that selects the panels. The same features that make a virtual-congress approach effective also expose such weaknesses as the potential for a bias that depends on the citizenship of the panelists who gather data for analysis. This increases the importance of including substantial proportions of non-US participants in all panels.

Multidisciplinary fields like materials science and engineering and immunology pose special challenges. For example, the immunology panel had to extract data from collaborative and international research; had to compare large enterprises with multiple smaller ones; and had to extract information on the specific field of immunology from related research fields in large, aggregate databases.

5.3 Other Observations about Benchmarking

The method by which the most important fields and subfields are identified is critical. For example, immunology is not considered a "discipline" in the traditional sense and does not have departmental status in most universities. The selection of subfields is a somewhat subjective process that might differ between one benchmarking exercise. Rather than being a drawback, however, such differences will reflect the continual shifting of the borders of modern fields. A field should be considered by the array of related domains between which investigators can move without leaving the realm of their expertise.

It is likely that benchmarking could be effective on a 3- or 5-year cycle because large fields of research change relatively slowly. Annual benchmarking probably would not be sensitive enough to reveal changes.

Our series of experiments has revealed that no benchmarking technique is sufficient by itself and that the utility of particular techniques varies by field. Therefore, each panel should use a variety of comparable qualitative and quantitative methods to afford cross-verification of results. The methods should be kept as independent as possible.

Because the accuracy of benchmarking depends heavily on panel members' personal knowledge of fields, panel members were more closely involved with the writing of the report than is frequently the case with committee-written reports.

Use of indicators that provide information on degree of uncertainty and reliability might enhance the presentation of the panel assessments of leadership status.

The extensive use of benchmarking would be enhanced by reliable, up-to-date information. The US field-specific data that are collected do not provide sufficient or timely information; non-US data are even more problematic.

A finding that the United States is the world leader in a research field might lead some to conclude that additional resources for that field are not warranted. This might or might not be the case. For example, the mathematics report indicates that the United States is the world leader in mathematics. If the mathematics community requests additional resources, some policymakers might question the request on the grounds that the United States is already the world leader in that field. That concern has been expressed in connection with the life sciences. However, as the mathematics panel indicated in its report, the United States could drop from being "the world leader" in mathematics research unless additional investments were made in some key subfields and unless more US students chose to enter the field. Thus, an assessment that the United States is the leader in a field does not necessarily imply that no additional resources are needed for the field.

6

CONCLUSION

On the basis of experiments in mathematics, immunology, and materials science and engineering, COSEPUP concludes that international benchmarking by panels of experts can provide an efficient and reasonably objective means of assessing the world leadership status of the United States in research fields.

The committee suggests further that such assessments can be completed in a timely fashion and at reasonable cost. Such a technique might be useful to policy-makers and federal agencies in evaluating their leadership status in research programs (for example, to comply with the GPRA).

In summary, benchmarking seems well suited to

- Assessing whether US researchers in a field are among or behind the world leaders or are the world leader and determining the leadership status of the United States in particular subfields.
- Identifying institutional and human-resource factors that are crucial to maintaining leadership status in a field.

APPENDIX A

COMMITTEE ON SCIENCE, ENGINEERING, AND PUBLIC POLICY: MEMBER AND STAFF BIOGRAPHICAL INFORMATION

Committee

Maxine F. Singer *(Chair),* president of the Carnegie Institute of Washington (Washington, DC), is an eminent biochemist whose wide-ranging research on RNA and DNA has greatly advanced scientific understanding of viral and human genes. Dr. Singer received her bachelor's degree from Swarthmore College (1952) and her PhD from Yale University (1957). She worked at the National Institutes of Health as a research biochemist in the National Institute of Arthritis and Metabolic Diseases until 1975, studying the synthesis and structure of RNA. In 1975, she moved to the National Cancer Institute. Her interest in primate DNA led to the discovery of a transposable element in the human genome. A member of the National Academy of Sciences and the Institute of Medicine, Dr. Singer has served on the editorial boards of several scientific journals and on the Governing Boards of Yale and the Weizmann Institute. She received the Distinguished Presidential Rank Award, the highest honor given to a civil servant, and the National Medal of Science in 1991.

Bruce M. Alberts, president of the National Academy of Sciences, is a respected biochemist recognized for his work in biochemistry and molecular biology. He is noted particularly for his extensive study of the protein complexes that allow chromosomes to be replicated, as required for a living cell to divide. He is a past chair of the Commission on Life Sciences and has served on the faculty of Princeton University and as vice chair and chair of the Department of Biochemistry and Biophysics of the University of California, San Francisco. Being committed to the improvement of science education, he has dedicated much of his time to education projects in San Francisco elementary schools.

Enriqueta C. Bond, president of the Burroughs Wellcome Fund, received her undergraduate degree in zoology and physiology from Wellesley College, a master's degree in biology and genetics from the University of Virginia, and a PhD in molecular biology and biochemical genetics from Georgetown University. She is a member of the American Association for the Advancement of Science, the American Society for Microbiology, and the American Public Health Association. She serves on the Board of Health Sciences Policy of the Institute of Medicine (IOM), the Board of the Society for the Advancement of Research on Women's Health, and the Board of the North Carolina Biotechnology Center. Dr. Bond was executive officer of IOM from 1989 to 1994.

Lewis M. Branscomb is the Aetna Professor of Public Policy and Corporate Management emeritus and former director of the Science, Technology, and Public Policy Program in the Center for Science and International Affairs at Harvard University's Kennedy School of Government. Dr. Branscomb graduated from Duke University in 1945, summa cum laude, and was awarded a PhD in physics by Harvard University in 1949. He has held teaching positions at the University of Maryland and the University of Colorado. He is a former president of the American Physical Society and of Sigma Xi, the Scientific Research Society. A research physicist at the National Bureau of Standards (now the National Institute of Standards and Technology) from 1951 to 1969, he was its director of NBS from 1969 to 1972. He is a member of the National Academy of Engineering, the National Academy of Sciences, and the National Academy of Public Administration. He serves on the Technology Assessment Advisory Committee to the Technology Assessment Board of the Congress. Dr. Branscomb is a former director of the IBM Europe, Middle East, Africa Corporation and of General Foods Corporation. He is a director of Mobil, MITRE, and the Lord Corporation and C. S. Draper Laboratories. He has written extensively on science and technology policy, comparative science and technology policy of different nations, information technology, management of technology, and atomic and molecular physics.

David R. Challoner is the director of the Institute for Science and Health Policy at the University of Florida. He served as vice president for health affairs and chairman of the Board of Directors of the Shands Health System at the University of Florida. He has held various professorships, including professor of internal medicine at the Indiana University School of Medicine, dean and professor of medicine at the St. Louis University School of Medicine, and professor of internal medicine at the University of Florida. His clinical specialty is internal medicine and endocrinology. Dr. Challoner is the foreign secretary of the Institute of Medicine. He was a member of the Institute's council, chair of its Membership Committee, and a member of COSEPUP. He also has served on the Advisory Committee to the Director of the National Institutes of Health, as president of the American Federation for Clinical Research, and as chair of the President's Committee on the National Medal of Science. He has received the Dr. William Beaumont Award of the American Medical Association.

Ellis B. Cowling is University Distinguished Professor At-Large and director of the Southern Oxidants Study at North Carolina State University. He earned a BS and an MS in wood technology at the State University of New York College of Forestry at Syracuse University, a PhD in plant pathology and biochemistry at the University of Wisconsin, and Filosofie Licensiat and Filosofie Doktor degrees in physiologic botany at the University of Uppsala in Sweden. He was elected to the National Academy of Sciences in 1973. His current research interests include changes in the chemical climate of industrial regions and their effects on aquatic and terrestrial ecosystems and the role of scientist and engineers in public-decision making. In 1995, he was appointed Visiting Eminent Scholar in the School of Earth and Atmospheric Sciences at Georgia Tech.

Peter Diamond is an Institute Professor and Professor of Economics at the Massachusetts Institute of Technology (MIT), where he has taught since 1966. He received his BA in mathematics from Yale University in 1960 and his PhD in economics from MIT in 1963. He has been president and chair of the National Academy of Social Insurance(NASI), president of the Econometric Society, and vice-president of the American Economic Association. He is a fellow of the American Academy of Arts and Sciences and a member of the National Academy of Sciences. He was the recipient of the 1980 Mahalanobis Memorial Award and the 1994 Nemmers Prize. He has written on public finance, social insurance, uncertainty and search theories, and macroeconomics.

Gerald P. Dinneen was foreign secretary of the National Academy of Engineering from 1988 to 1995. He was previously vice president of science and technology at Honeywell Corporation and from 1977 to 1981 was the assistant secretary of defense and principal deputy under secretary of defense for research and engineering. He has had a long affiliation with the Massachusetts Institute of Technology (MIT) having joined the MIT Lincoln Laboratory in Lexington, MA, in 1955. He advanced through many positions to become Director in 1970 to 1977 and professor of electrical engineering in 1971-1981. He was elected to the National Academy of Engineering in 1975 and serves on many advisory committees and boards in the National Research Council and in government. He has been elected to the Engineering Academy of Japan, the Swiss Academy of Technological Sciences, and the Royal Academy of Engineering of the UK.

Mildred S. Dresselhaus is the Institute Professor of Electrical Engineering and Physics at the Massachusetts Institute of Technology, where she directed the Center for Materials Science and Engineering. She has been active in the study of a wide array of problems in the physics of solids and the structure and properties of carbon fibers, fullerenes and carbon nanotubes. She was awarded the National Medal of Science in November 1990 and was elected to the National Academy of Engineering in 1974 and to the National Academy of Sciences (NAS) in 1985. She has been a member of the Councils of both academies and of the Governing Board of the National Research Council, treasurer of NAS, and president of the American Physical Society and of the American Association for the Advancement of Science.

James J. Duderstadt is president emeritus and University Professor of Science and Engineering at the University of Michigan. He received his BA from Yale University in 1964 and his doctorate in engineering science and physics from the California Institute of Technology in 1967. He joined the faculty of the University of Michigan in 1968 and has served as professor of nuclear engineering, dean of the College of Engineering, provost, vice president for academic affairs, and president from 1984 to 1996. He received the National Medal of Technology for exemplary service to the nation, the E.O. Lawrence Award for excellence in nuclear research, and the Arthur Holly Compton Prize for outstanding teaching. He has served as chair of the National Science Board, chair of the Board of Directors of the Big Ten Athletic Conference, and chair of the Executive Board of the University of Michigan's hospitals. He serves as a director of the Unisys Corporation and CMS Energy Corporation. He has been a member of the National Academy of Engineering since 1987.

Alexander H. Flax, was Home Secretary of the National Academy of Engineering from 1984-1992. He received his bachelor's degree in aeronautical engineering from New York University in 1940 and a PhD in physics from the University of Buffalo in 1957. Previously, he was president of the Institute for Defense Analyses, chief scientist of the Air Force, assistant secretary of the Air Force for research and development, director of the National Reconnaissance Office, and vice-president and technical director of the Cornell Aeronautical Laboratory. He was a US national delegate to the NATO Advisory Group for Aerospace R&D from 1969 to1987.

Marye Anne Fox, a chemist and member of the National Academy of Sciences (NAS), is North Carolina State University's 12th chancellor. Previously, she was M. June and J. Virgil Waggoner Regents Chair in Chemistry and vice president for research at the University of Texas at Austin. Her research interests include physical organic chemistry, organic photochemistry, organic electrochemistry, chemical reactivity in non-homogeneous systems, heterogeneous photocatalysis, and electronic transfer in anisotropic macromolecular arrays. She serves on the Council of NAS, its Executive Committee, and the Committee on Science and Education Policy. After Senate confirmation in 1990 of her nomination to the National Science Board, she served as its vice chairman (1994-1996) and chaired its Committee on Programs and Plans (1991-1994). She serves on the Texas Governor's Science and Technology Council, has chaired the Chemistry Section of the American Association for the Advancement of Science, and advises its Center for Science, Technology, and the Congress. She has served on advisory panels for the Army, the Department of Energy, the National Science Foundation, and the National Institutes of Health. She has served on 14 editorial boards, including a stint as associate editor of the *Journal of the American Chemical Society*. She serves on boards of the Texas Environmental Defense Fund, the Texas Agribusiness Council, the Texas Food and Fiber Commission, W.R. Grace, and Oak Ridge Associated Universities.

Ralph E. Gomory has been president of the Alfred P. Sloan Foundation since 1989. After being a Higgins Lecturer and assistant professor at Princeton, he joined IBM in 1959, and became vice president in 1973, and was senior vice president for science and technology in 1985-1989. A member of both the National Academy of Sciences and the National Academy of Engineering, he has received the Lanchester Prize in 1963, the John von Neumann Theory Prize in 1984, the IEEE Engineering Leadership Recognition Award in 1988, the National Medal of Science in 1988, the Arthur M. Bueche Award of the National Academy of Engineering in 1993, and the Heinz Award for Technology, the Economy and Employment in 1998. He was named to the President's Council of Advisors on Science and Technology in 1990 and served to March 1993.

Ruby P. Hearn is senior vice president of the Robert Wood Johnson Foundation, which has awarded over $2billion in grant funds since its inception as a national philanthropy in 1972. As a member of the executive management team, she participates in strategic program planning with the president and executive vice president and serves as a special adviser to the president and as the foundation's liaison in the nonprofit community. Dr. Hearn has had the major responsibility for oversight and program development of initiatives in maternal, infant, and child health, AIDS, substance abuse, and minority-group medical education. She received her MS and PhD in biophysics from Yale University and is a graduate of Skidmore College. She is a fellow of theYale Corporation. She served on the Executive Committee of the Board of Directors for the 1995 Special Olympics World Summer Games in Connecticut. She is a member of the Institute of Medicine and its' Council, COSEPUP, the Board of Directors of the Council on Foundations, the Science Board of the Food and Drug Administration, and the Advisory Committee to the Director of the National Institutes of Health.

Brigid L. M. Hogan is an investigator with the Howard Hughes Medical Institute and Hortense B. Ingram Professor in the Department of Cell Biology at Vanderbilt University School of Medicine. She obtained her PhD from Cambridge University, England, and carried out postdoctoral training at the Massachusetts Institute of Technology. Before moving to the United States, she was head of the Laboratory of Molecular Embryology, first at the Imperial Cancer Research Fund and then at the National Institute of Medical Research in London. Dr. Hogan is a member of the European Molecular Biology Organization and was recently elected to the Institute of Medicine.

Samuel Preston became dean of the University of Pennsylvania School of Arts and Sciences in January 1998 and has been a faculty member in sociology since 1979. He is a scholar of population studies with expertise in technical demography and the analysis of mortality and family structure. He has served twice as chair of the Department of Sociology, three times as chair of the Graduate Group in Demography, and as director of the Population Studies Center and Population Aging Research Center. Dr. Preston is a member of the National Academy of Sciences, the Institute of Medicine, the American Academy of Arts and Sciences, the American Association for the Advancement of Science, and the American Philosophical Society. Earlier in his career, he served as a faculty member at the University of California, Berkeley and the University of Washington. He was acting chief of the Population Trends and Structure Section of the UN Populations Division from 1977 to 1979. Dr. Preston holds a BA from Amherst College and a PhD in economics from Princeton.

Kenneth I. Shine is president of the Institute of Medicine and professor of medicine emeritus at the University of California, Los Angeles (UCLA) School of Medicine. He is UCLA School of Medicine's immediate past dean and provost for medical services. He has also been director of the Coronary Care Unit, chief of the Cardiology Division, and chair of the Department of Medicine at the UCLA School of Medicine. He has served as chair of the Council of Deans of the Association of American Medical Colleges, and was president of the American Heart Association. His research interests include metabolic events in the heart muscle, the relation of behavior to heart disease, and emergency medicine.

Morris Tanenbaum was the vice Chair of the Board and chief financial officer of AT&T from 1988 to 1991. He began his career at Bell Telephone Labs on the technical staff and held various positions at Western Electric Company, including vice president of the Engineering Division and vice president of manufacturing, before returning to Bell Labs in 1975 as executive vice president. In 1978, he became president of New Jersey Bell Telephone Company. He returned to AT&T as executive vice president for Corporate Affairs and planning in 1980, and became the first Chair and CEO of AT&T Communications in 1984. He was vice president of the National Academy of Engineering until June 1998.

Irving L. Weissman is Karele and Avice Beekhuis Professor of Cancer Biology, professor of pathology and professor of developmental biology at Stanford University School of Medicine. Dr. Weissman was a member of the Scientific Advisory Board of Amgen (1981-1989), DNAX (1981-1992), T-Cell Sciences (1988-1992). He was a co-founder of SyStemix and was chairman of its Scientific Advisory Board and a member of its Board of Directors in 1988-1997. His main research interests are hematopoietic stem cells, lymphocyte differentiation, lymphocyte homing receptors, and phylogeny of the immune system.

Sheila E. Widnall served as secretary of the Air Force from 1993 to 1997. Dr. Widnall received her BSc (1960), MS (1961), and ScD (1964) in aeronautics and astronautics from the Massachusetts Institute of Technology (MIT). She was appointed Abbey Rockefeller Mauze Professor of Aeronautics and Astronautics at MIT in 1986 and served as associate provost of MIT from 1992-1993.

William Julius Wilson is the Lewis P. and Linda L. Geyer University Professor at Harvard University. He was formerly Lucy Flower University Professor of Sociology and Public Policy at the University of Chicago. He received the National Medal of Science in 1998. He is a member of the National Academy of Sciences, the American Academy of Arts and Sciences, and the National Academy of Education; a former member of the President's Committee on the National Medal of Science; and a past president of both the American Sociological Association and the Consortium of Social Science Associations.

William A. Wulf is president of the National Academy of Engineering (NAE). He was AT&T Professor of Engineering and Applied Science at the University of Virginia. He has served as assistant director of the National Science Foundation, chairman and CEO of Tartan Laboratories, Inc., and professor of computer science at Carnegie Mellon University. He has been a member of NAE since 1993 and serves as chair of the Computer Science and Telecommunications Board.

Staff

Richard E. Bissell is executive director of the Policy Division of the National Academy of Sciences and Director of COSEPUP. He took up his current position in June 1998. Most recently, he served as coordinator of the Interim Secretariat of the World Commission on Dams (1997-1998) and as a member and chair of the Inspection Panel at the World Bank (1994-1997). He worked closely with the National Academy of Sciences during his tenure in senior positions at the Agency for International Development (1986-1993), as head of the Bureau of Science and Technology, and of the Bureau of Program and Policy Coordination. He has published widely in political economy and has taught at Georgetown University and the University of Pennsylvania. He received his BA from Stanford University (1968) and his MA and PhD from Tufts University (1970, 1973).

Deborah D. Stine is associate director of COSEPUP, director of the Office of Special Projects, and director of the National Academies Christine Mirzayan Internship Program. She has worked on various projects in the National Academies since 1989. She received a National Research Council group award for her first study for COSEPUP, on policy implications of greenhouse warming, and a Commission on Life Sciences staff citation for her work in risk assessment and management. She has also worked on studies on research and the government performance and results act, science and technology centers, risk assessment and management, graduate education, responsible conduct of research, careers in science and engineering, and environmental issues. She holds a bachelor's degree in mechanical and environmental engineering from the University of California, Irvine; a master's degree in business administration; and a PhD in public administration, specializing in policy analysis, from the American University. Before coming to the National Academies, she was a mathematician for the Air Force, an air-pollution engineer for the state of Texas, and an air-issues manager for the Chemical Manufacturers Association.

APPENDIX B

EXCERPTS FROM NATIONAL ACADEMIES REPORTS

APPENDIX B-1

EXCERPT FROM: SCIENCE, TECHNOLOGY AND THE FEDERAL GOVERNMENT: NATIONAL GOALS FOR A NEW ERA

Chapter 3
National Goals for Science

Science and technology are closely linked to important national objectives in such areas as economic growth, health care, national security, and environmental protection. These linkages implicitly raise the question of what goals the nation should have for science and technology and how those goals should be reflected in levels of investment, organizational arrangements, and subjects for research and development. In this chapter, we examine these issues for science; in the next chapter, we do so for technology.

In setting national goals for science, several observations must be kept in mind. The first is that it has proved impossible to predict reliably which areas of science will ultimately contribute to important new technologies. History is rich in examples of scientific research that have led to practical applications in areas far removed from the original work. Fundamental research on electromagnetism contributed directly to the development of modern communications. Investigations in solid-state physics enabled the invention of the transistor. The recombinant DNA technology that led to the biotechnology industry arose from studies of unusual enzymes in bacteria. Mathematics, often regarded as highly abstract, is at the core of applications as diverse as aircraft design, computing, and predictions of climate change.

The second observation involves the importance of research for which applications are not yet known. For investigators who do such fundamental research, primarily in universities, the original motivation is seldom to develop new applications; rather, it is the desire to discover and to understand natural processes. Nevertheless, this motivation—enlarging the store of human knowledge—in the end brings advances and applications that cannot be made any other way. A substantial redirection of such fundamental research toward goal-directed work would reduce the potential for advances of economic and social importance without necessarily leading to solutions for the problems being addressed.

A third observation concerns the cultural significance of science. The application of modern scientific research, beginning in the seventeenth century, is one of the most profound events in human history. Modern research has done more than change the material circumstances of our lives. It has changed our ideas about ourselves and our place in the universe, about human history and the human future.

A final observation has to do with leadership. The great modern expansion of scientific knowledge has led to changes that have been of enormous benefit to humanity. Adjusting to these changes has in some cases proved disruptive to society. Nonetheless, leadership in science has become one of the defining characteristics of great nations. The United States has risen to a position of global prominence in part through its strengths in science and technology. Those strengths can continue to contribute greatly to U.S. leadership.

In light of the above observations, we believe that the federal government, in partnership with the private sector and with other levels of government, should adopt explicit national goals for science. Our first recommendation is:

> **The United States should be among the world leaders in all major areas of science.**

"Major areas" refers to broad disciplines of science (such as biology, physics, mathematics, chemistry, earth science, and astronomy) and to their major subdisciplines (such as the neurosciences, condensed-matter physics, and seismology). "Among the world leaders" means that the United States should have capabilities and infrastructures of support that are not exceeded elsewhere. Of course, there will be specific areas or skills in which other nations lead the world. But in considering the major subdisciplines in which such areas belong, the United States should meet world standards.

There are several rationales for this goal:

- *Excellence.* When U.S. researchers are working at world levels in all disciplines, they can bring the best available knowledge to bear on problems related to national objectives, even if that knowledge appears unexpectedly in a field not traditionally linked to that objective. For example, by being among the world leaders in the areas of virology, immunology, and molecular biology, U.S. researchers were able quickly to devise a test for AIDS antibodies that helped ensure the safety of the blood supply.

- *Receptiveness.* By being among the world leaders in all disciplines, U.S. researchers can quickly recognize, extend, and utilize significant research results that occur elsewhere. For example, high-temperature superconductivity was discovered in Switzerland, but U.S. researchers were able to repeat and extend these findings within a matter of days.
- *Education.* Only by working in the presence of world leaders can students in American colleges and universities prepare to become leaders themselves and to extend and apply the frontiers of knowledge.
- *Personnel.* Maintaining excellence in a field is the best way to attract the brightest young students to that field and thus ensure its continuing excellence.

In general, being among the leaders in each area of science means that U.S. scientists understand and participate in expanding the frontier of human knowledge. The United States could not have been the early home of the semiconductor industry without having been among the world leaders in solid-state physics. It could not have been the home of the emerging biotechnology industry without having been a world leader in molecular biology.

In addition to being among the world leaders in all areas of science, the United States will wish to excel in certain areas on a national level. Therefore, the committee's second recommendation is:

The United States should maintain clear leadership in some major areas of science.

The rationales for maintaining a clear lead in selected areas of science go beyond those listed above. Among the criteria that would call for clear leadership in a field are the following:

- The field is demonstrably and tightly coupled to national objectives that can be met only if U.S. research performers are clear leaders. For example, the field of condensed-matter physics drives technological advances in such industrial sectors as microelectronics, advanced materials, and sensors.
- The field so captures the imagination that it is of broad interest to society. An example in astronomy is the recent detection of differences in the cosmic background radiation related to the creation of the universe.

- The field affects other areas of science disproportionately and therefore has a multiplicative effect on other scientific advances, especially those where clear leadership is the objective. For example, molecular biology is critical to advances in health care, biotechnology, agriculture, and industrial processes.

The selection of those fields in which the United States wishes to maintain clear leadership will be made by government decisionmakers with appropriate advice from various interested groups. These decisions must be fully informed by the comparative assessments of different scientific fields, discussed below, and by the extent to which different fields meet the criteria for clear leadership. These decisions thus differ in character from decisions about the most promising directions for research *within* an area of science, which are made most effectively by researchers themselves and should be insulated from the political process.

Implications of the Performance Goals

The federal government needs a better way to gauge the overall health of research—as a whole and in its parts—and to determine whether it is adequately supporting broad national objectives. Such indicators as dollars spent or numbers of scientists supported are, by themselves, inadequate. Nor can such indicators determine the adequacy of overall funding or the appropriate distribution of funds among different fields.

The committee believes that comparative international assessments of scientific accomplishment are a better yardstick for policy decisions. This concept, like others in this report, has been discussed in theory and applied in specific cases, but the committee believes that it now deserves a central place in national considerations of science policy. To this end, we have developed it in much greater detail.

The committee believes that it is feasible to monitor U.S. performance with field-by-field peer assessments. Researchers in many fields have, in recent years, identified research opportunities and even set funding priorities. The processes that they have used could be adapted for more general application.

The committee recommends the establishment of independent panels consisting of researchers who work in a field, individuals who work in closely related fields, and research "users" who follow the field closely. Some of these individuals should be outstanding foreign scientists in the field being examined.

The panels would assess the performance of U.S. research scientists in a given field and compare it with performance of researchers in other nations. To do this, the panels might, for example, judge where the most exciting and promising ideas are emerging, consider where the best new talent is locating, and examine the comparative capabilities of research facilities or equipment. The panels could identify key factors enhancing—or blocking—performance within different fields and project trends into the future. They would assess both the internal performance of the field and its relationship to other fields of science. Finally, the panels could recommend actions for both the performers and the supporters of research (see box).

Quantitative measurements, such as movements of individuals, literature citation counts, and quantity and quality of instrumentation, would be important tools for the panels. But the most valuable contribution by the panels would be the qualitative judgments of panel members working in each field. Scientists immersed in a particular field are best qualified to appraise the true quality of the work being done, to identify the most promising and exciting advances, and to project the status of the field into the future. The participation of representatives from other fields of science and from the users of research would help to allay concerns that the panels' recommendations are self-serving.

We believe that assessments of fields will prove useful in the alloca-

RESULTS OF THE PANELS' DELIBERATIONS

The independent panels of researchers and research users, drawing on their assessments of the comparative performance of U.S. researchers, could make several kinds of reports to the broader research community and to the federal government. Here are three hypothetical outcomes:

1. In one area of science, an assessment panel might find that the U.S. research effort is not at world levels of performance. The panel would diagnose the reasons for the deficiency—perhaps inferior facilities or a shortage of qualified young researchers. It could then translate its findings into proposals for the funders, performers, and users of research that would help bring U.S. performance up to world standards.
2. In another field, an assessment panel might find that U.S. performance is at world levels and does not seem to be in danger of falling behind. In this case, the group could recommend actions that would keep the United States among the world leaders.
3. In yet another area, an assessment panel might conclude that the United States leads the rest of the world. If that area is not one in which the United States should maintain clear leadership, the panel might recommend reductions in funding, which could then be applied to areas requiring additional support.

tion of resources both within and among fields. Within fields, the assessments will help identify the key factors affecting the comparative performance of its researchers. The assessments will be much more useful than current budgetary criteria for analyzing issues such as the adequacy of the infrastructure and the optimal number of students entering the field. By providing long-term perspectives, the assessments will increase the predictability and stability that are essential to a focused and sustained effort in science.

Assessments of research performance would also help resolve debates over the support of megaprojects. The costs of a megaproject would be assessed in terms of the performance goals. If the area of the megaproject was one in which the United States chose only to be among the leaders, participation would almost certainly depend on international collaboration and cost sharing. For an area where clear leadership is justified, the United States might choose to pursue a megaproject even without international partners. One could envision this process being applied, for instance, to the Human Genome Project, the Superconducting Supercollider, NASA's Mission to Planet Earth, fusion reactors, synchrotron radiation light sources, and so on.

The goals that we recommend also have implications for the research infrastructure. Meeting the goals requires that appropriate elements of the U.S. research infrastructure remain second to none. Educational institutions are essential to this infrastructure; only by providing the finest instruction in mathematics and science can the United States produce world class young scientists and engineers.

The committee believes that these goals can be met within the existing overall federal R&D budget. First, because of its traditionally strong support for science, the United States is already a leader in most areas of science. Second, through the application of the goals outlined above, the federal government is likely to find that the United States has clear leadership in some areas of science in which we need only to be among the leaders; funds can be redistributed accordingly. Third, relatively minor reallocations of the federal government's R&D budget, which now exceeds $70 billion per year, could have a major effect on the research portion of the budget.

A New Framework for Funding

The performance goals stated above would provide the basis for a new approach to designing and enacting federal research budgets.[3] Today, the federal R&D budget emerges from a process that is only loosely coordinated. Each federal agency supports research in pursuit of its individual mission, and research is often a relatively small part of that mission.

Guided by the performance goals of being among the world leaders in all areas of science and maintaining a clear lead in some, the Executive Branch and Congress could take a more coherent approach to setting R&D budgets. In the Executive Branch, the assessments of scientific fields could guide initiatives designed to achieve specific scientific or technological goals. In the past few years, initial steps in this direction have been taken by the Federal Coordinating Council for Science, Engineering, and Technology under the Office of Science and Technology Policy. If a major field of science were found to be behind world standards, the Executive Branch could boost funding across all the agencies that support research in the field.

When the budget reaches Capitol Hill, the House and the Senate would conduct comprehensive reviews of the proposals for science and technology before disaggregating the budget for agency-by-agency examination. Limited versions of such reviews now take place in both houses, but they need to be structured so that their results have more impact on the decisions of individual appropriations subcommittees.

APPENDIX B-2

EXCERPT FROM: ALLOCATING FEDERAL FUNDS FOR SCIENCE AND TECHNOLOGY

RECOMMENDATION 4. The President and Congress should ensure that the FS&T budget is sufficient to allow the United States to achieve preeminence in a select number of fields and to perform at a world-class level in the other major fields.[8]

The pool of approximately $35 billion to $40 billion in annual public support for FS&T is large and diverse. The committee believes that it is possible within that budget to reduce some programs, eliminate others, increase support of high-opportunity fields, and restrain federal spending—all while maintaining our nation's tradition of excellence in science and technology. To continue as a world leader, the United States should strive for clear leadership in the most promising areas of science and technology and those deemed most important to our national goals. In other major fields, the United States should perform on a par with other nations so that it is "poised to pounce" if future discoveries increase the importance of one of these fields. If the nation sets priorities in this way (see bulleted items below) and uses them in conjunction with the FS&T budget process, the result will be better decisions about reallocating and restructuring the U.S. research and development enterprise, preserving its core strengths, and positioning it well for strong future performance.

The international comparisons needed to assess U.S. achievement of its goals for leadership in research and development should be conducted by panels of the nation's leading experts under White House

Box 1.4 Evaluating FS&T Opportunities and Making International Comparisons: How It Might Work

Every five years, panels are convened to evaluate the fields in each major area of science and technology (e.g., physics, biology, electrical engineering), their standing in the world, and the resources needed to reach and maintain world-class position. Evaluation focuses on outputs, such as important discoveries, and also on certain benchmarks of best practice, such as number of scientists and engineers and their training or the current state of the laboratories and research facilities. To avoid conflicts of interest, at least half of the panel will include a few nonscientists plus experts from fields outside but related to the fields being evaluated. The panel will also include specialists in the evaluated fields who are recruited from the United States and foreign countries. If any field within a major area is performing below world standards but is judged to be a national priority, the panel will recommend that its budget be augmented or other changes made to bring it up to par. At the same time, the panel will identify the other fields with declining scientific opportunities and obsolete federal missions from which resources should be reallocated. Opportunities for international cost-sharing will be examined to achieve optimal use of federal funds devoted to science and technology.

Evaluations will be commissioned by the National Science and Technology Council or its equivalent. The selection of fields for clear U.S. leadership from among those recommended by the panels will be made by the President and presidential advisors as part of the budget process. As an example, an extract of the President's budget message might read: "I propose that the United States need not be so far ahead in experimental particle physics, but should operate at world levels, in this case by contributing to construction of the particle accelerator in Geneva, sponsored by the CERN, and funding the participation of U.S. scientists in its design and research. On the advice of my Council of Advisors on Science and Technology, I propose that the United States should remain clearly preeminent in the molecular biology of plants and animals for the following reasons. . . . Accordingly, I will include the necessary additional funds in the FS&T budgets of the National Institutes of Health, the Department of Agriculture, and the National Science Foundation to achieve this goal. . . ."

auspices. Reallocation decisions should be made with the advice and guidance of these expert panels, capable of determining the appropriate scope of the fields to assess and to judge the international stature of U.S. efforts in each field (see Box 1.4 for a discussion of how international comparisons might work). These panels would recommend to the President, his advisors, and Congress:

- Which fields must attain or maintain preeminence, based on goals such as economic importance, national security, unusual opportunity for significant discoveries, global resource or environmental issues, control of disease, mitigation of natural disasters, food production, a presidential initiative (such as human spaceflight), or an unanticipated crisis;
- Which fields require increases in funding, changes in direction, restructuring, or other actions to achieve these goals; and
- Which fields have excess capacity (e.g., are producing too many new investigators, have more laboratories or facilities than needed) relative to national needs and international benchmarks.

The committee believes that designing the budget process so as to secure an FS&T budget sufficient to ensure preeminence in select fields and world status in others will allow the United States to maintain continued world leadership. The FS&T budget process must be coupled to systematic review of investments by the nation's best scientific and technical experts, reporting to the highest reaches of government, to produce an appropriately balanced mix of activities. The committee emphasizes that wise federal investments will lead to the creation of new wealth in the future to an even greater extent than they have in the past. As a result, these investments will help reduce the federal deficit in the long run. After a period of budget constraints, reconfiguration, and adjustment, national needs may justify increased investments in FS&T.

APPENDIX C

WORKSHOP ON INTERNATIONAL BENCHMARKING OF US RESEARCH

Summary

COSEPUP held a 1-day workshop on June 16, 1999, to discuss the methodology and utility to policy-makers of its three benchmarking experiments with invited guests from federal agencies, Congress, universities, and other institutions.

During the morning session, leaders of the three benchmarking experiments (Peter Lax, mathematics; Arden Bement, materials science and engineering; and Irv Weissman, immunology) summarized the work of their panels. During the afternoon, discussants provided comments on the utility and methodology of the benchmarking experiments.

Highlights

- Panel leaders started out as skeptics, but came to believe that the process of benchmarking is feasible, quick, and accurate.
- Benchmarking can produce a rapid, broadly accurate "snapshot" of a field. With greater rigor and the generation of relevant local data, it can probably be applied with specificity as well, for example, to evaluate particular agency programs.
- Through the use of a "virtual congress", it is possible to get 80% of the value in 20% of the time.
- Benchmarking produces data, not policy decisions.
- Benchmarking should use both qualitative tools (the "virtual congress") and quantitative tools (citation and publication analysis, prizes, and presentations).
- Because of the statistical risks in using small samples for rapid assessments, all available tools should be used.

• The benchmarking model might be adapted at the agency level to evaluate research programs and instruct advisory committees. Agencies would need to determine their own benchmarks, such as comparison with other US scientists or programs.

• Two independent benchmarking experiments in mathematics—despite dissimilar panels, mandates, and leadership—produced similar results, lending credibility to the technique.

• Experts on several panels suggested that a diminished flow of foreign-born scientists and engineers to the United States could weaken the research enterprise in coming years. This underscores the importance of drawing more American students into science.

• In seeking to explain the overall dominance of US research, participants pointed to diversity, flexibility, research-based graduate education, a balanced research portfolio, national imperatives, and a favorable innovation climate.

• It was suggested that an accumulation of benchmarking exercises might lead to better understanding of the factors that yield research excellence and might better educate the public about the value of research.

Report

The following excerpts and quotations are offered to summarize the issues discussed.

Was the Exercise Successful?

The benchmarking chairs were initially skeptical that the study could be conducted. For example, Arden Bement, chair of the materials science and engineering study, said:

> "I started out as a skeptic, but became more of a believer that 1) it's possible to do, 2) it can be done within a short time, 3) the committee got a lot out of it. It's not perfect. I think we got 80-90% of what we looked for, within the scope we chose to probe. You can get 80% of the value in 20% of the time. After we'd finished, my thinking on it was much clearer than before."

Comments received by Dr. Bement after the release of the report were positive. The concerns were focused on sub-subfields not addressed, such as research areas important to industry (for example, corrosion).

In addition, some surprises were illuminated as a result of the analysis. For example, Peter Lax, chair of the mathematics panel, indicated that although he knew that many leading mathematicians were from abroad, he was surprised to find out the degree to which US leadership depended on non-US talent in the United States.

Methodology

All panels used the same general methods: a combination of qualitative judgment by experts and quantitative tools (measures of publications, citations, prizes, and speakers at international meetings). Within each experiment, all assessment methods—qualitative and quantitative—gave similar results.

The use of a "virtual congress", or "reputation survey", was found to be effective for determining the relative standing of a nation's research. In immunology, for example, the panel created a virtual congress by identifying and polling international leaders in major subfields. These leaders were asked to imagine that they were about to organize an international meeting in their particular sub-subfield and to furnish a list of five to 20 potential speakers for that meeting. The identities of the speakers were used to create a "snapshot" of international leadership.

An advantage of the virtual congress is that results are available quickly. "The results of [a first quick] go-round were virtually the same as the final result, which was that about half of the best [immunology] scientists are in the US."

The virtual congress identified sub-subfields of great current interest but did not perform in-depth analysis. In materials research, "the utility [of benchmarking] depends on what kind of assessment you want. We were talking to experts and asking them for hot topics, not looking at whole portfolios."

Concerns About Methodology

Several participants voiced concerns about the virtual congress. One was that it was small and therefore in danger of being biased. In addition, there was no consensus on how many foreign members a panel should have (immunology had three, two of whom have since moved to the United States) and what effect foreign membership (or lack of it) had on decisions. Several people suggested that half the members of each benchmarking panel should be non-American scientists or engineers.

Small panel sizes might be unavoidable if the goal is to poll the most talented leaders: "In asking our questions we quickly started hitting the same people." A general concern was that the method of selecting people can somewhat bias the outcome by promoting an accepted "party line" and by including fewer people who are inclined to follow less-popular directions. However, panel leaders said that the advantages of reaching those most knowledgeable and active in a field outweighed those drawbacks.

Several other concerns about methodology were noted:

- Panel members might use different modes of polling, variable sample sizes, and subjective criteria in assessing leadership.

- The use of citations might sometimes be skewed by clustering in some areas, self-referencing, and cross-referencing among certain authors.
- For benchmarking reports to be of greatest use, results should be described as fully as possible. For example, if 50% of a field's leaders are in the United States, are 49% in a single other country, or do 10 countries each have 5% of the leaders?
- There is a suggestion that the United States is "number one" in most fields because it spends the most money on research. Another standard of performance might be "research productivity." That is, if the United States funds 60% of the research in a field, it should attain at least 60% of the leadership in that field. However, another nation might fund 10% of the research in the field and be able to attain 25% of the leadership in that field. The second country might be considered more effective at attaining leadership for its level of funding.
- When the United States is dominant even with respect to whole regions (such as Europe), it is difficult to make country-by-country comparisons.

One person suggested that the validity of benchmarking should be tested by repeating an experiment with a different panel or by using two expert panels simultaneously. Such testing, however, might be hindered by the small number of experts and might not produce new results. Two independent tests were done in mathematics, but they produced essentially the same conclusions, even though the two panels were quite different in composition. One panel consisted mostly of mathematicians (it included eight American mathematicians), the other mostly of non-mathematicians (it included two American mathematicians); one was asked to draw conclusions, the other was asked <u>not</u> to draw conclusions.

Journal-publication and citation analysis suffered some shortcomings. In immunology, journal analysis was generally limited to the largest international journals, which tend to be American or English (such as *Nature, Science, Cell,* and *Blood*). In mathematics, the use of only mathematics journals eliminated many fields of which mathematics is a part (geophysics, probability, and mathematical biology); for this reason, the judgment of experts familiar with sub-subfields was especially valuable.

One participant urged greater uniformity of methods so that results would be comparable and easier to use by agencies or Congressional committees. A panel chair responded that the cost of seeking uniform comparability might not add value compared with small, frequent, "snapshot" assessments. He suggested trying an experiment with more-rigorous methods to see whether they produced different results.

Social-science terms are sometimes used in the reports without sufficient rigor. For example, the concepts of "leader" and "fast follower" should be defined uniformly among reports and connected to earlier COSEPUP reports. Is it "almost as good" to be a fast follower as to be a leader? When is maintaining leadership worth the investment required?

On Being a "Fast Follower"

Panel members suggested that in some fields, being a "fast follower" can be a good strategy. It is unrealistic to think that any country can be the leader in every field or subfield of science or engineering. However, a country can support scientists and programs that are "among the leaders" in every field. The advantage of being among the leaders is that a nation can react quickly to new discoveries or fields that are suddenly hot. The main text uses the example of high-temperature superconductivity, a sub-subfield in which the United States was not the leader but was able to move rapidly because US researchers were among the leaders in related subfields or sub-subfields.

The Question of Timing

Workshop members discussed several aspects of timing. The choice of year and even of decade can alter results; had the mathematics benchmarking been done 10 years ago, the Soviet Union would have figured much larger.

There was also a concern about frequency. It is not clear, one participant said, whether benchmarking would lose its impact if done every 5 years, because the same leadership status and issues might be repeated. Another commented that regular evaluations would provide the opportunity to determine whether situations had improved or remained the same.

Several agency representatives raised the issue of "old data", that is, the National Science Foundation (NSF) and science and engineering indicators come out every 2 years and require an additional 2 years for validation, and this creates at least a 4-year lag. Panel members responded that an assessment by a virtual congress can be done relatively quickly, using data that are current.

Quantitative and Qualitative Benchmarking

Several participants asked whether benchmarking could use more rigorous, quantitative measurements. In response, COSEPUP members said that quantitative analysis (of numbers of publications, citations, patents, and so on) is helpful in assessing some research programs or projects, especially when the goal of the research is an incremental improvement or achievement of a known goal. But expert judgment is required to analyze the relative importance of various journals, citations, and patents—the tools of quantitative analysis.

Moreover, quantitative tools offer little information about important aspects of research programs. The current judgment of practicing researchers, managers, policy experts, and users of research is needed to answer such questions as where the most promising ideas are emerging, what locations are being chosen by the best new scientific talent, and what the comparative quality of research facilities in different areas is.

General Comments on the Science and Engineering System

Diversity and flexibility are essential qualities of the American system that allow a rapid upshift from fast follower to leader. For example, NSF recognizes the value of these qualities by not penalizing researchers who divert funds to areas of new or emerging importance. One panel member noted that in the last dozen years, none of the important discoveries made in his laboratory [IW] had been anticipated in a grant proposal. He argued against hierarchies or planned strategies: one cannot know where important new areas might emerge.

One panelist said that the strength in immunology depends heavily on pluralism, decentralized funding, and entrepreneurship: "We are so entrepreneurial that everyone is their own principal investigator. There are many points of light."

In discussing potential threats to the research system, a number of participants mentioned the tendency of industry and government to emphasize applied, results-driven research at the expense of high-risk, high-return basic research. A healthy research enterprise requires a balanced portfolio of research. As one panelist said, "one can't cultivate only the fruit of the tree."

Factors That Lead to Strength in Science and Engineering Research

Human resources, breadth, and motivation were mentioned:

- Human resources: There was concern that not enough American students are staying in mathematics and there is little understanding of why they don't. One participant said, "it depends on your point of view. US students might make the logical choice to go into computer science instead."
- Breadth: A "healthy" field requires not just a few stars, but broad strength—both creators (leaders) and innovators (followers).
- Motivation: The truly successful are those most motivated. A person needs intelligence, but what defines the real achievers is "everything else". Motivation might not be apparent at a very early age—a strong argument against preselecting early achievers and in favor of early mentoring of all students.

Benchmarking, with properly defined terms, might show the factors that have brought fields to their present positions. By capturing the key factors, the process could elucidate serious issues worthy of case studies and could lead to a better understanding of the science and engineering system.

The process might also be a useful tool to demonstrate how fields evolve, what brings success, and how funding is decided. Many people and agencies have little understanding of the scientific enterprise. An accumulation of GPRA reports could be a valuable educational and public-policy instrument.

Funding alone does not determine leadership. National imperatives and the ability to capitalize on research have also been major factors.

Foreign Scientists and Engineers

On the benchmarking panels, strong-minded and independent foreign participants are crucial to successful evaluations.

One of the surprises to the panels was the number of foreign-born scientists and engineers in the United States. "We knew that many American mathematicians had come from abroad, but there were more than we thought." The United States attracts the best PhDs and postdoctoral students from abroad, adding to the excellence of US education. Some participants called the United States "very dependent on them, maybe overly so."

The US is attractive to foreign scientists and engineers in part because of the flexibility of the system. "In the [former] Soviet Union or Europe, a professor gets a chair and stays there. Here they move around," increasing the diversity of the US research enterprise.

At the same time, it is hard to attract US students to enter mathematics and science, partly because of the strong economy and the low perceived economic value of advanced degrees in some fields. As foreign universities catch up, the need to attract more of the best US students will probably grow.

Agency Responses

Several representatives of federal agencies that support research gave valuable perspectives on the concept of benchmarking and on COSEPUP's experimental efforts.

Some agencies might be able to adapt for their own use the benchmarking model, which they can then apply with greater specificity and rigor.

Some features of the experiments might be useful to agencies for instructing their own advisory committees.

A representative of the National Institutes of Health (NIH) found the experiment in immunology "very useful; it validates the picture of US research with a large number of experts from the area. It will allow us to authenticate these results to a greater degree than we now can. It adds strength to our request for a balance of funds in NIH. We are familiar with this approach, without asking for quantitative data to back it up. It helps us in requesting additional resources when this is necessary. The data on subfields gives us guidance to go and explore those areas further. We do not find it useful to make comparisons with other countries, but we do want to know about areas that are not being developed and probably should be."

Summary of Suggestions

In summary, workshop participants suggested the following actions to improve the benchmarking process:

- Conduct an additional set of experimental benchmarking studies of agency research programs within the context of GPRA.
- Increase the number of non-US researchers on each panel to 50%.
- Evaluate fields of interest to industry, as well as to the federal government.
- Invite representatives of federal agencies and national laboratories to discuss topics for benchmarking and to participate in benchmarking itself.
- Augment analysis of research with economic and market data.
- Determine international leadership status for both industrial research and academic research.
- Focus not only on the standing of individual researchers but also on the research establishment as a whole.
- Evaluate the research community not only for leadership status, but also for its ability to be a "fast follower".
- Assess the nation's ability to capitalize on the results of research.
- Evaluate the research standing not only for the present, but also in relation to previous assessment(s), to gauge progress.
- Analyze how funding is allocated among research, education, and facilities.
- Analyze the relative roles of researchers in universities, federally funded facilities, and industry.
- Provide the degree of detail and rigor required by policy-makers to make funding decisions based on the report.
- Include international leadership status in the charge to existing advisory committees that review federal programs.
- Provide clearer definition of "leadership" in a field.
- Make methods as comparable as possible.

- Conduct a study twice, with different committees, and compare results.
- Undertake more careful, in-depth studies of issues identified through benchmarking.

ATTACHMENT 1

INTERNATIONAL BENCHMARKING OF US MATHEMATICS RESEARCH

Panel on International Benchmarking of
US Mathematics Research

Committee on Science, Engineering, and Public Policy

International Benchmarking of US Mathematics Research

Panel Members

PETER D. LAX *(Chair)*, Professor of Mathematics and Director, Courant Mathematics and Computing Laboratory, New York University, New York, NY

MICHAEL F. ATIYAH, Master, Trinity College, Cambridge, England

SPENCER J. BLOCH, Professor, Department of Mathematics, University of Chicago, Chicago, IL

JOSEPH B. KELLER, Professor, Departments of Mathematics and Mechanical Engineering, Stanford University, Stanford, CA

JACQUES-LOUIS LIONS, President, French Academy of Sciences and Professor, College de France, Paris, France

YURI I. MANIN, Director, Max Planck Institut fur Mathematik, Bonn, Germany

RUDOLPH A. MARCUS, A.A. Noyes Professor of Chemistry, California Institute of Technology, Pasadena, CA

GARY C. McDONALD, Head, Operations Research Department, GM Research and Development Center, Warren, MI

CATHLEEN S. MORAWETZ, Professor Emeritus, Courant Institute of Mathematical Sciences, New York University, New York, NY

PETER SARNAK, Chair, Department of Mathematics, Princeton University, Princeton, NJ

I.M. SINGER, Institute Professor, Massachusetts Institute of Technology, Cambridge, MA

MARGARET H. WRIGHT, Distinguished Member of Technical Staff, Bell Laboratories, Lucent Technologies, Murray Hill, NJ

Project Staff

DEBORAH D. STINE, Study Director

PATRICK P. SEVCIK, Research Associate

JOHN R. TUCKER, Senior Program Officer

NORMAN GROSSBLATT, Editor

Mathematics Benchmarking Guidance Group

MARYE ANNE FOX *(Chair)*, Vice President for Research, University of Texas at Austin, Austin TX

PHILLIP A. GRIFFITHS, Director, Institute for Advanced Study, Princeton, NJ

DANIEL KLEPPNER, Professor of Physics, Massachusetts Institute of Technology, Cambridge, MA

AVNER FRIEDMAN, Director, Institute for Mathematics and its Applications, University of Minnesota, Minneapolis, MN

SAMUEL WINOGRAD, IBM Fellow, International Business Machines Corporation, Thomas J. Watson Research Center, Yorktown Heights, NY

ROBERT MacPHERSON, Professor, School of Mathematics, Institute for Advanced Study, Princeton, NJ

Staff:

LAWRENCE E. McCRAY, Executive Director, COSEPUP

DEBORAH D. STINE, Associate Director, COSEPUP

JOHN R. TUCKER, Director, Board on Mathematical Sciences

NORMAN METZGER, Executive Director, Commission on Physical Sciences, Mathematics, and Applications

CONTENTS

Figures and Tables

Report

EXECUTIVE SUMMARY

The United States is clearly preeminent in mathematics today. The field is thriving in terms of both quality and opportunities. Not only are there stellar researchers in all fields at American institutions, but they are backed by a broad and active research community. Mathematical research in the United States has many links with science, engineering, and technology and is broadening its contacts with education at all levels. But this position of eminence is fragile. Increasing demands are placing a strain on the mathematics community.

In making judgments about mathematics, the International Benchmarking of US Mathematics Research Panel kept these points in mind:

- Mathematics is the language and tool of most of the sciences.
- Mathematical results often have a long life.
- Mathematical research is conducted on a very broad front, and seemingly disjointed branches often turn out to be intimately related.
- Ideas of abstract mathematics often are crucial ingredients in practical applications.
- Mathematics is one of the pillars of education in kindergarten, elementary school, high-school, and college.

The present strength in US mathematics is due to:

- Continued attractiveness of the United States to talented people around the world.

- A strong system of graduate education.
- Diversity and flexibility of the US research enterprise.
- Sustained funding for research from universities and the federal government.

The United States continues to attract some of the best graduate students and postdoctoral fellows from all over the world; a substantial portion of active research mathematicians now in the United States come from outside the United States. But we are in danger of losing our preeminent position if we do not face some critical issues and challenges. Some critical issues and challenges must be faced:

- US leadership in mathematics rests on the health of research universities, which today are experiencing severe financial pressure and conflicting demands.
- The United States is not taking sufficient advantage of its native mathematical talent: while graduate enrollment from abroad thrives, the number of American students applying to graduate school in mathematics is diminishing.
- Serious thought is needed about how to make better connections between mathematics and other fields, because mathematics is crucial in much interdisciplinary research.
- US industry has reduced its commitment to long-range research in mathematics.

1

BACKGROUND

In 1993, the Committee on Science, Engineering, and Public Policy (COSEPUP) of the National Academy of Sciences, the National Academy of Engineering, and the Institute of Medicine issued the report *Science, Technology, and the Federal Government: National Goals for a New Era*. This report recommended that the United States be among the world leaders in all major fields of science and maintain clear leadership in selected fields. A similar recommendation was made in a later National Research Council (NRC) report, *Allocating Federal Funds for Science and Technology*, published in 1995—that the United States "strive for clear leadership in the most promising areas of science and technology and those deemed most important to our national goals."

Both reports stated that quantitative measures, such as dollars spent and number of scientists supported, were inadequate indicators of leadership and that policy decisions about programmatic issues or resource allocation would be better informed by comparative international assessments. Independent field-specific panels were suggested as the best means for obtaining such evaluations. Each panel would consist of researchers in the particular field, researchers in closely related fields, and research users who follow the field, and each panel would include researchers from outside the United States.

In late 1996, COSEPUP began an experimental study of the effectiveness and outcome of such panels. The present report—an evaluation of US research in mathematics—was prepared by the first panel and will be followed by studies in materials science and immunology. Each panel has been asked to address the following questions:

- What is the position of the United States in research in the field relative to that in other regions or countries?
- What key factors influence relative US performance in the field?
- On the basis of current trends in the United States and abroad, what will be the relative US position in the near term and the longer term?

2

SCOPE AND NATURE OF THE PANEL'S EVALUATION

Mathematics is the most formal and rigorous of the sciences, but there is no universally accepted definition of *mathematics* (and the panel has not attempted to formulate one). In this report, to conform with the 1992 report *Educating Mathematical Scientists: Doctoral Study and the Postdoctoral Experience in the United States*, published by the NRC Board on Mathematical Sciences, *mathematics* broadly includes pure mathematics, applied mathematics, statistics and probability, operations research, and scientific computing.

Mathematics has several properties that complicate its assessment:

- Mathematical research generally has a particularly long "shelf life": large parts of it do not become obsolete. Not infrequently, a much earlier result or insight—even from a previous century—is suddenly the key to solving a modern problem.
- Seemingly diverse subfields of mathematics often form unexpected links.
- Throughout science and engineering, mathematics provides a universal language, tools for analysis, abstractions to guide understanding, and methods for solving problems. Consequently, mathematical research has a tightly coupled, two-way connection with other fields: mathematical discoveries influence research in other fields, and developments in other fields provide new problems for mathematicians to study. However, the contributions of mathematical research are often not labeled explicitly as such.
- Mathematical training is a central part of the education of all US citizens, from kindergarten through high school and college.

Because of the first two properties, we have not defined and separately assessed subfields of mathematics, but rather have focused on mathematics as a whole. Because of the third and fourth properties, our evaluation reaches beyond mathematics to fields and activities where mathematics research has a direct and visible impact. We also recognize that important mathematical research is conducted by people whose affiliations and titles do not explicitly identify them as mathematicians; some of the difficulties associated with this phenomenon are described in SIAM (1995).

US mathematics research was defined by the panel as research in the mathematical sciences conducted by residents of the United States working in US institutions.

The panel would like to mention 5 key caveats with respect to its analysis. First, following its charge from COSEPUP, this report is based on the qualitative judgments of the panel members informed by both their own knowledge and the sparse quantitative data available. The panel has attempted to be as fair and impartial as possible, balancing the points of view of US academic mathematical researchers with views of leading mathematicians from outside the United States, nonmathematical US researchers, and industrial researchers. In addition, the panel was specifically charged not to make recommendations. With more time and effort, additional opinions and data could have been collected, but such efforts would have been expensive relative to the additional guidance obtained.

Second, given the diversity of mathematical sciences, no panel could represent all the subfields of mathematics. The panel has done its best to review all subfields, but some are undoubtedly better analyzed than others. The findings and conclusions here do not apply uniformly to all areas of the mathematical sciences. For example, statistics has enjoyed much stronger employment prospects than most other areas. Thus, some variation in interpretation of these results is required when focusing on specific areas.

Third, many statements in this report are not based on numerical data, mainly because of the paucity of statistics that allow meaningful comparisons among countries. For example, undergraduate and PhD degrees in the United States are not directly comparable with all similarly labeled degrees in other industrialized countries. Even when quantitative information is available, sometimes it contains so many ambiguities that we were reluctant to rely on it. To understand the position in other countries when suitable data were unavailable or unclear, the panel relied on the informed judgments of panel members and colleagues from outside the United States.

Fourth, a substantial amount of mathematics is carried out by people bearing other labels: physicists, chemists, electrical engineers, economists, computer scientists, and statisticians. Some Nobel prizes in these fields have been awarded for mathematical work.

Fifth, had this report been written 7 years ago, the Soviet Union would have loomed large as a competitor of the United States. The collapse of the Soviet Union changed that; Russia is in disarray on all fronts, infrastructure is collapsing, and many of the best former-Soviet mathematicians have found employment abroad, particularly in the United States. At present, mathematics in Russia and Ukraine is a shadow of its former self.

3

RELATIVE POSITION OF US RESEARCH IN MATHEMATICS

3.1 The Discipline

3.1.1 Leadership

Our first means of evaluation was an ad hoc survey. The panel divided the mathematical sciences into 19 subfields corresponding roughly to the classification used by the International Mathematical Union, and each panel member was assigned a set of subfields in which he or she was knowledgeable. For every subfield, the assigned panel members identified a reasonable number (between 5 and 10) of internationally recognized leaders in several countries. These experts were asked to draw up a list of about 10 speakers (10 is the average number of speakers on a given subfield at an international congress on mathematics), without regard to nationality, for a hypothetical international mathematics congress. Speakers were to be selected because their research was at the leading edge and driving the subfield. Our intent was not to identify the most-famous or best-established people, but rather those whose research was the most important at the time. The demographic results of this informal process were remarkably uniform in each subfield and across subfields. In 17 of the 19 subfields, at least half those named are citizens or permanent residents of the United States. In the other two, about 40% of those named work in the United States.

A second means of evaluating research leadership was to examine the lists of winners of major prizes in mathematics. This measure can be criticized on several grounds: far fewer prizes are awarded in mathematics than in many other sciences, prizes do not uniformly cover all subfields of mathematics, and prizes awarded on an international basis are invariably subject to (unstated) requirements that they not be dominated by a single country. Nonetheless, the panel felt that the clear leadership of US mathematics was demonstrated by looking at two of the most prestigious prizes in mathematics: the Fields medal and the Wolf prize. The Fields medal is presented every 4 years at the International Congress of Mathematicians (ICM) and is by tradition given to mathematicians younger than 40. Of the 38 Fields medals awarded so far, 14 (about 37%) went to people in the United States; more than 40% of Fields medal winners are now working in the United States. The Wolf prizes have been given annually since 1978 for outstanding achievement in physics, chemistry, medicine, agriculture, the arts, and mathematics. More than half (15 of 28) of the recipients of the Wolf prize in mathematics, which is not restricted by age of the recipient, now live in the United States. Moreover, although there is no Nobel prize in mathematics, the 1994 Nobel prize in economics was shared by a US mathematician, John F. Nash.

A third indicator is the US representation among the plenary speakers at two large and prestigious international mathematics meetings. The ICM is typically attended by about 4,000 mathematical researchers. ICM speakers are chosen by distinguished committees with some attention to balanced geographic distribution among the speakers. At the last ICM, in 1994, 8 of 16 (50%) of the 1-hour plenary speakers were from the United States. In 1990 (Kyoto), 9 of 15 (60%) were American, and in 1986 (Berkeley), 8 of 16 (50%). For the International Congress on Industrial and Applied Mathematics (ICIAM), attended by about 2,500 mathematical researchers, plenary speakers are chosen by a committee representing mathematics societies from 12 countries or regions, and substantial attention is paid to balancing the plenary speakers among those countries. In the 1995 ICIAM, 7 of 20 (35%) of the invited plenary speakers were from the United States, and in 1991, 6 of 20 (30%).

3.1.2 Depth

The three indicators just described reflect clear US leadership based on the accomplishments of a relatively small number of stellar mathematicians. We felt that we should also assess the more-robust measure of *depth* in research leadership; US leadership would be fragile if it depended on the location of a few individuals. We believe that the United States has substantial depth in all subfields of mathematics, on the basis of the following observations:

- In the United States, 183 institutions award PhDs in the mathematical sciences. The 20 or so top-ranked mathematics departments in this group are comparable in research excellence with those at the best universities anywhere in the world.

- As shown in figure 1, US mathematicians consistently produced about 40% of the total research publications in mathematics from 1981 to 1993.
- The American Mathematical Society (AMS), an organization of researchers in the mathematical sciences, has 30,000 members, including 22,000 US members. Non-US citizens join AMS through reciprocity agreements. The Society for Industrial and Applied Mathematics (SIAM), an organization of researchers in applied mathematics and scientific computing, has 6,400 US members in a total membership of 9,000. Annual attendance by US residents at the joint research conferences held by AMS and the Mathematical Association of America (MAA) is about 3,500. Annual attendance by US residents at SIAM research conferences is about 2,300. The meetings and publications of these societies play a major role in disseminating mathematical ideas. Membership in the societies is a rough measure of sustained interest.

3.2 Mathematics in a Broader Context

As mentioned in section 2, the quality of mathematical research can partly be measured by its effects on closely related activities. We consider four: scientific and engineering research, industry, government, and education. It is difficult to carry out this analysis, because it is hard to document which mathematics-related activities are conducted by mathematicians and which by people trained in other scientific and engineering fields.

3.2.1 Science and Engineering

Numerous studies have documented in great detail the strong connections of mathematical sciences research with the physical,

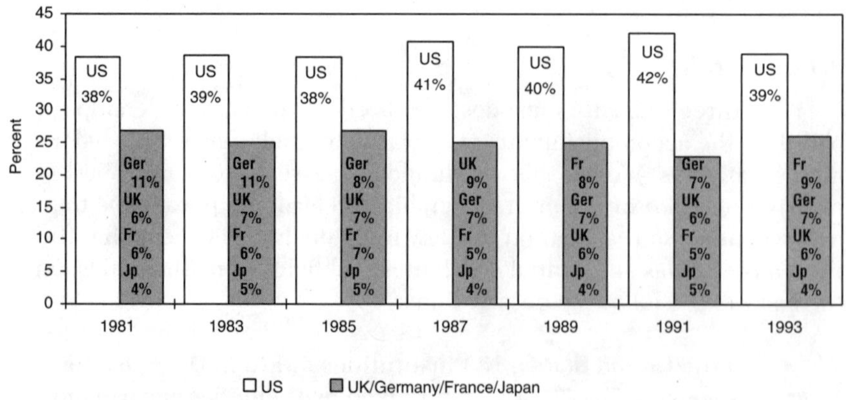

FIGURE 1 Percentage of mathematics-research papers published by US authors. Source: NSB 1996, appendix table 5-31.

biologic, and social sciences, engineering, and medicine. We list a small subset of diverse recent instances in which US mathematical research is closely linked with other fields. That a broad spectrum of mathematics contributed to these examples emphasizes the *unity* of the mathematical sciences (see section 2).

- Physics has been the science closest to mathematics for the longest time, and their closeness continues today. Many of history's most famous scientists worked in both physics and mathematics (from Newton, Euler, Gauss, Lagrange, Poisson, Kelvin, Maxwell, Poincaré, and Rayleigh to Einstein, Weyl, von Neumann, and Witten). Mathematical physics is extremely active in the United States; many questions of common interest in mathematics and physics arise from quantum mechanics and field theory, general relativity, fluid and multiphase flow, electromagnetic theory, and materials science. Semiconductor modeling, thin films, and signal transmission in optical fibers are three special areas of high mathematical content (see, for example, NRC 1993).
- Biology, physiology, and theoretical and computational chemistry are adopting mathematical approaches, and some eminent US researchers in these fields actively collaborate with mathematicians (NRC 1995a, b). *Mathematical Challenges from Theoretical/Computational Chemistry* (NRC 1995a) describes recent successes, such as the development of commercial products from quantitative structure-activity relationships and the insights into molecular structure gained from group theory and topology. Materials scientists and mathematicians have increasingly been forming research partnerships. A 1995 Minerals, Metals, and Materials Society-SIAM workshop on modeling microstructural evolution (Chen and others 1996) produced 70 papers by US mathematician, physicist, and materials scientist coauthors. A joint 1996 initiative of the National Science Foundation (NSF) and the Defense Advanced Research Projects Agency (DARPA) in materials science attracted numerous proposals involving research from all subfields of mathematics.
- Computer science traditionally relies on particular branches of mathematics, certainly logic but also combinatorics and number theory. Recently, the astonishing growth of traffic on the Internet has led to new applications in queuing theory, discrete mathematics, combinatorial optimization, and protocol verification. The report *Cryptography's Role in Securing the Information Society* (NRC 1996a) discusses the need for mathematical research in, for example, number theory and logic to develop and analyze cryptographic techniques guaranteed to remain secure when faced with continuing gains in raw computing power.
- Imaging has relied on and inspired new mathematics for over 20 years. The basis of the CAT-Scan, the Radon transform, was first described nearly 80 years ago. Mathematical research today in, for

example, deblurring and real-time detection of anomalies links directly with medical applications. The report *Mathematics and Physics of Emerging Biomedical Imaging* (NRC 1996b) states that "many of the envisioned innovations in medical imaging are fundamentally dependent on the mathematical sciences." The mathematics of imaging is also important in astronomy, biology, geosciences, weather, and cartography.

- Engineering, in all its fields, uses sophisticated mathematics to formulate, analyze, and solve problems, particularly those in which early prototyping and experimentation are too expensive or too risky. The 1995 symposium on "Frontiers of Engineering" held at the National Academy of Engineering highlighted four topics of research in engineering—biotechnology, design and manufacturing, environmental engineering, and information technology; US mathematicians have been active contributors to all four (NAE 1996; SIAM 1995).
- In meteorology, biotechnology, and other "grand-challenge" problems, mathematicians and scientists in other fields have had major successes. The nature of such interdisciplinary collaborations is discussed in two studies—1 by COSEPUP (1996) and the other by the National Academy of Public Administration (NAPA 1995).

That list demonstrates the success of US mathematics research in taking part in research in science, engineering, and medicine. Much of it has been achieved through computing, both locally and through supercomputer centers. There was no feasible way for the panel to determine the interactions of mathematics with science and engineering in other countries. Panel members from outside the United States and our sampling of reports from other countries confirm that all industrialized nations are vigorously encouraging interdisciplinary research in which mathematics plays a part (see section 5.2).

3.2.2 Industry

Many large companies in the United States that rely on technical innovation support research and development laboratories that employ PhD mathematicians. In 1995, 21% of doctoral mathematical scientists in the US workforce were employed in private industry—a proportion that has steadily increased over the years (see table B-1). In 1975, 11% of PhD mathematicians worked for industry; in 1985, 19%. That trend might indicate that the flexibility of mathematics PhD programs is increasing. A few US industrial research laboratories—Bell Labs, IBM Research, and General Electric—have been world-famous for their research in mathematics and in other sciences. Major US companies, such as AT&T, Boeing, and General Motors, have maintained active, high-quality groups of research mathematicians. However, it is often difficult to identify mathematical research in medium- or small-scale industrial settings because the organizational structures are cross-disciplinary.

In many widely publicized instances, mathematical research has made substantial contributions to US industry successes. The aerodynamic design of the Boeing 777 was accomplished by computing airflow, pressure, temperature in the exterior of the proposed design; research on numerical methods, adaptive grid generation, and optimization was crucial. The visualization system allowed thousands of scientists, engineers, and customers to work together (Council on Competitiveness 1996). Mathematics was also central to the remarkable animation in the 1995 Disney film *Toy Story* (SIAM 1996). Another example is the soliton, discovered around 1965 by the mathematicians N.J. Kruskal and M.D. Zabusky, that is now poised to play a key role in the transmission of signals in optical fibers. The mathematical concept of wavelets is an increasingly important tool for the storage and recovery of information, for instance, fingerprints.

Mathematical research has played well-documented roles in many other fields of industry that have strengthened the economic position of the United States, particularly in the automotive, pharmaceutical, communication, and computer industries (NRC 1991). Within the last few years, there has been a great deal of interest by banks and investment houses in employing mathematicians to use mathematical techniques to model and analyze financial trends (COSEPUP 1995); financial mathematics depends on a number of recent discoveries and techniques found in mathematics. Attempts to assess precisely the contributions of mathematical research to industrial problems are hindered by the blurring of disciplinary boundaries throughout most of industry (SIAM 1995).

Especially since the 1980s, US academic mathematicians have created programs and organizations, partially funded by industry, that are designed to involve mathematics research with industrial problems. Examples include the Institute for Mathematics and its Applications at the University of Minnesota and several programs at public and private universities.

Outside the United States, strong connections exist between research mathematics and industry. In France, a rapid development immediately followed World War II in the fields of applied mathematics and scientific computing; many strong researchers took jobs in industry. In Germany, recently created institutes combine academic and industrial partners, for example, in aerospace and automotive manufacturing. Several institutions in the UK support research interactions between mathematicians and industry. Examples are the Rolls-Royce Readership in Computational Fluid Dynamics at Oxford, the Newton Institute at Cambridge, and the Basic Research Institute in the Mathematical

Sciences at Bristol (founded with partial funding from the UK branch of Hewlett-Packard).

3.2.3 Government Laboratories and Agencies

During and after World War II, the US government established laboratories in which research in mathematics (and many other disciplines) was supported. In 1993, government laboratories employed 4.5% of the mathematical-sciences PhDs in the United States (NSF 1996a, table 20). Even with the cuts in defense spending that began in the early 1990s, US government labs have maintained substantial investments in mathematical research. Mathematical research is an integral part of the mission of federal laboratories, such as those of the Department of Energy at Los Alamos, Livermore, Berkeley, Argonne, and Oak Ridge; several Department of Defense laboratories; of the National Security Agency; of the National Institute of Standards and Technology; of the National Center for Atmospheric Research; of the National Aeronautics and Space Administration; of the National Oceanic and Atmospheric Administration; and other agencies such as the Bureau of the Census of the Department of Commerce.

The position of mathematics research conducted in government laboratories varies in other industrialized countries. In the United Kingdom, for example, until the 1980s, the missions of several government laboratories (such as the National Physical Laboratory and AERE Harwell) included basic research, but this has now been de-emphasized or eliminated. There apparently remain a large number of research mathematicians in the Defense Research Agency, but their numbers are classified. Major decreases in defense spending in the UK have led to concomitant reductions in mathematical research in the associated laboratories. Nonmilitary research in government laboratories has been cut; several nondefense laboratories that employed research mathematicians have been privatized and have moved toward commercial, short-term activities, rather than long-term research.

In France, government research laboratories were established in 1939, and mathematics plays a prominent role in several of them (for example, Institut National de Recherche en Informatique et en Automatique (INRIA) and Institute de Hauts Etudes (IHES)). We found no data specifically about mathematics; national laboratories account for 22% of all research in France, and half of French scientists work full-time on research in laboratories run by government agencies. About 20% of all scientists and engineers work in government laboratories (NSF 1996c, p. 36).

International comparisons are difficult because of differences in organizational structures; for example, many government laboratories in France are integrated within universities.

3.2.4 Mathematics Education

Of all the sciences, mathematics is most closely scrutinized for its function in education, largely because mathematical skills are seen as a key indicator of the scientific and technologic development of the citizenry. The panel stresses the need for a healthy relationship between mathematical research and mathematics education. To bring modern, simpler ways of looking at material and to incorporate appropriate recent research into graduate and undergraduate education in mathematics and other fields, the role of active researchers is crucial. Research experiences that encourage mathematical skills and innovation are a growing part of US undergraduate mathematics education.

Research mathematicians from all sectors of the US higher-education system, including the most-prestigious mathematics departments, are increasingly involved in improving the teaching of mathematics, at every level of education, to both specialists and nonspecialists. "Service teaching" of undergraduate mathematics to nonmajors is the responsibility of mathematics departments in all US universities, and many leading US academic mathematicians regularly teach elementary courses for mathematics nonmajors. There have always been divided views on this. There was a period when many nonmathematics departments taught these courses; however, when research universities expanded, these departments were happy to shift this responsibility to mathematics departments. Now the reverse is occurring due to the general shrinking of research universities. The best solution appears to be better teaching by mathematicians operating in full cooperation with the departments concerned. New methods based on that idea have led to major improvements in calculus teaching.

Policy-making and professional organizations of US research mathematicians are deeply involved in education. The NRC Board on Mathematical Sciences, the NRC Mathematical Sciences Education Board, AMS, MAA, and SIAM regularly initiate studies and publish reports about various aspects of mathematics education. Collaboration between high schools and research mathematicians is rare but increasing. However, despite the increased efforts of many mathematicians, we have a long way to go, compared to many other countries, in teaching mathematics effectively in kindergarten through high school, as illustrated in the recent evaluation of 8th-grade mathematics instruction in the Third International Mathematics and Science Study (TIMSS) (ED 1996).

It is beyond the scope of this report to discuss the enormous task of improving the mathematical level of the general population. The contribution of research mathematics to this task in only a small though essential part. However, the level of mathematics education at the K-12 level affects the research community in two ways:

- Inadequate high-school education makes it more difficult for college and university instructors to maintain standards and create an intellectually challenging curriculum.
- Because of the lack of interesting exposure to mathematics, fewer students are interested in studying mathematics or pursuing mathematics as a career.

Mathematics education is a major concern in all other countries of which the panel is aware, but we have only anecdotal data about how research mathematicians elsewhere are involved in education. There is a general concern in Europe (except in France) that the quality of education is going down. Japan, according to a recent report, has an outstanding mathematics curriculum (Askey 1993).

The early recognition and training of the mathematically talented is traditional in Russia, France, Hungary, Romania, and Poland. We are happy to report that this tradition is gaining ground also in the United States, through special high schools, special publications for the young, correspondence courses, statewide contests, national and international olympiads, intense summer programs, and so on.

4

FACTORS THAT INFLUENCED PAST US PERFORMANCE IN MATHEMATICS

We have identified four leading influences on the success of US mathematical research: attractiveness to foreign talent, quality and structure of graduate education, diversity of the mathematical research enterprise, and financial support for research and infrastructure.

4.1 Attractiveness to Talent from Outside the United States

A policy of welcoming distinguished scientists as citizens or permanent residents has enabled the United States to attract from abroad many of the world's best senior mathematicians and promising young mathematicians. Leading scientists, including mathematicians, fled to the United States from the Nazis during 1933-1945 and were followed by a second flood after World War II, from 1945-1955. This concentration of immigrants raised the level of US mathematics to the top. Substantial increases in the number of outstanding mathematicians immigrating to the United States—for example, from China and the former Soviet Union—have also occurred more recently. The pattern of assimilating top talent from all nations outside the United States has been consistent and striking.

America has long been viewed as the "promised land" of freedom, wealth, and opportunity. In addition, mathematicians were drawn to the United States for several practical reasons, discussed below—more and better jobs, high salaries, funding for research, and greater mobility than in any other country. On the last issue, for example, European professors tend not to move once they secure a chair. In France, professors migrate toward Paris. In contrast, it is not uncommon for many of the best US professors to change jobs repeatedly.

4.2 Quality and Structure of Graduate Education in Mathematics

Because the brightest students want to study with the best people, the presence, described above, of leading mathematicians at universities throughout the United States has been a major factor in the visibility and appeal of US graduate education since the end of World War II, when the "GI Bill" enabled poor but talented students to take advantage of educational opportunities.

US graduate education in the sciences, mathematics, and engineering has been concurrently boosted by two other influences. After the launch of Sputnik in 1957, the United States adopted various national policies that strongly encouraged the study of mathematics, science, and engineering from elementary through graduate school. The large number of "baby boom" undergraduates entering colleges and universities in the 1960s and 1970s led to substantial expansion in mathematics departments and graduate programs throughout the United States.

One structural aspect of US graduate education in mathematics stands out in comparison with other countries: the much lower level of specialization required to enter a graduate program. For example, it is possible to enter a US PhD program in mathematics without an undergraduate degree in mathematics; such late shifts of major are extremely rare in other countries. This "late start" feature increases flexibility and choice for prospective students. By the time US students begin their dissertation research, they are typically as well prepared as their counterparts elsewhere, but possibly older.

Strong US graduate programs in mathematics have been able to attract top-quality students not only in the United States but also from abroad, showing again the great appeal of the US mathematics environment to foreign talent. A well-known example is the enrollment in the 1980s of a large number of brilliant graduate students from China, Taiwan, Hong Kong, and the USSR. Many of them then remained in the United States after receiving the PhD, and some outstanding mathematicians in the United States today belong to this group.

4.3 Diversity of the US Research Enterprise

Before World War II, only a handful of US research universities were distinguished in mathematics; today, at least 2 dozen have uniformly high-quality faculty across most subfields of mathematics, and many more have stellar researchers in particular subfields. Rather than being dominated by a few institutions or individuals, this diffuse structure allows a wide range of mathematicians from across the entire country—and with their institutions—to excel. The diffusion of talent is strengthened by a level of professional mobility that is unmatched elsewhere.

Four independent research institutes in the mathematical sciences contribute to the quality of US mathematics. The Institute for Advanced Study, founded in the 1930s, has long been a major force in pure mathematics, drawing talented people from around the world. The Center for Discrete Mathematics and Theoretical Computer Science at Rutgers, the Mathematical Sciences Research Institute at the University of California, Berkeley and the Institute for Mathematics and its Applications at Minneapolis—all funded largely by the National Science Foundation (NSF)—have been created since 1980. A somewhat different model is the Courant Institute which is integrated with the mathematics and computer science departments at New York University. Those four and the National Institute of Statistical Sciences have increased awareness of research accomplishments, brought leading and junior researchers together, provided support for postdoctoral students, and created ties between different subfields of mathematics—for example, geometry and mathematical physics—and between mathematics and industry. There are many analogous institutes abroad (for example, the Max Planck Sonderforschung and the Oberwolfach in Germany, the Euler Institute in Russia, the Mittag-Leffler Institute in Sweden, the Newton Institute at Cambridge, and the IHES in France).

Since World War II, mathematics in US universities has branched out and expanded into new fields. The invention and development of electronic computers provided a stimulus for mathematics throughout the world. The United States pioneered in the use of computers, thanks partly to the leadership of von Neumann, and to the success of the semiconductor, software, and computer industry. Computing in England began during the war, primarily from the work of Alan Turing. The rest of the industrial world is catching up, but the United States still dominates.

The field of computer science was spawned jointly by mathematics and electrical engineering, and many parts of computer science remain closely linked with mathematics. Consistent and dramatic increases in computing power have encouraged mathematicians to tackle long-standing problems, formulate and solve new problems, devise new numerical methods, and produce software. Graduate programs that combine mathematics with various scientific fields have been initiated. Thus, the growing roles of mathematics in science, engineering, and medicine are formally recognized and encouraged in many places, as discussed in section 3.2.1. The emerging role of mathematics in business, finance, and modern management has also been spawned by new mathematical methods and greater computer power.

4.4 Adequate Funding

The three factors already mentioned explicitly rely on sustained funding for mathematical research, which comes from various sources, both public and private. Funding for individual faculty members gives them time to concentrate on research, and it supports graduate students; funding for conferences, workshops, summer schools, and other infrastructure facilitates interactions that are central to a thriving mathematical research community. These have greatly increased the exchange of information through personal discussion in mathematical research over the last two decades.

The predominant element in funding of United States mathematics research has been the strong commitment to intellectual excellence by private and public universities. To preserve and build research quality, universities have been willing to expend financial resources to hire and support the world's best mathematicians, as noted above in the discussion of the diversity of US mathematical research. That has occurred since World War II. Before then, mathematics professors were expected to focus on teaching, and research was considered an attractive sideline except at a few elite institutions, such as Harvard, Yale, Princeton, and Berkeley. Today, many institutions still focus on teaching, but almost 50 focus on research as well.

The second-most important element is support by the federal government. Federal funding for mathematics began during World War II when the United States Office of Scientific Research and Development recruited mathematicians and other scientists to work on applied problems of military significance. Soon after the war, the US government established the Office of Naval Research, the NSF, and other agencies to support scientific research. At the same time, existing government research laboratories were enlarged, and new ones were created. Today, the leading agencies supporting basic research in mathematics are NSF, the Department of Defense, the Department of Energy, and the National Security Agency. In the decentralized American system, federal funds have played a vital role in promoting communication and enabling institutions to maintain world-class research by individual faculty and small research groups.

In addition, faculty at state universities receive research funding from the states, and some private universities offer extra research support for faculty who do not receive federal funding. Several private foundations—such as Sloan, Guggenheim, Ford, and Packard—offer awards for junior faculty, senior-faculty sabbaticals, and special projects.

5

CURRENT TRENDS

It is obvious, but should nonetheless be emphasized, that broad economic and political trends in the United States affect mathematics research. This report cannot possibly address all the complex and controversial issues concerning, for example, optimal mechanisms for federal and industrial support of research, the proper role of research universities, and the pressures of international competition. Many reports on these topics have been produced during the last few years by COSEPUP, the Office of Science and Technology Policy, the National Science Board (NSB), the Industrial Research Institute, the Council on Competitiveness, and others.

With specific reference to mathematics, the US preeminence in mathematical research, described in section 3. has been attained in large part because of the factors listed in section 4. However, unemployment among recent PhDs has created tremendous stress on US mathematics during the 1990s. In this section, we identify a variety of current trends—positive and negative—that are affecting or are likely to affect the relative position of US mathematical research in scientific accomplishments and development of the knowledge base.

5.1 Vitality of the Mathematical Sciences

The panel wishes to emphasize that US mathematical research is thriving in both quality and opportunities. Many new subfields of mathematics have been developed, and some major long-standing problems have been solved, thereby opening new avenues for solving other problems (as did Wiles's proof of Fermat's last theorem). New methods of solution have been introduced, new connections between different fields have been discovered, and new ways to apply mathematics in science and engineering have been found. Computing has transformed, and will continue to transform, all subfields of mathematics. Mathematicians worldwide express a similar level of enthusiasm for their field.

5.2 Interdisciplinary Research

Although most people agree that interdisciplinary science should be encouraged, there is no universally accepted strategy for doing so. The relative effectiveness of different approaches will be understood only after more experience is gained. In the meantime, US research mathematicians are continuing to play active—in many instances, leading—roles in interdisciplinary research. To name just one topic of current interest, mathematical research will be crucial in making sense of massive data sets (NRC 1996c), particularly when data-gathering happens adaptively in real time; lack of progress in this arena is a recognized impediment to progress in biology, medicine, astronomy, physics, and geosciences.

US universities and federal funding agencies are trying to create programs that encourage mathematicians in all subfields to create links with other disciplines. In times of tight budgets, however, it is difficult to justify moving money away from already-squeezed disciplinary research programs that have consistently produced outstanding results. Anecdotal evidence suggests that interdisciplinary programs, especially those perceived as "risky," are struggling to adapt within existing structures.

The United States is not alone in attempting to devise policies that support interdisciplinary science and engineering despite budget pressures. The European Union and other Europe-wide programs explicitly seek to improve scientific cooperation among the countries of the region. To prepare scientists for international work, increasing mobility of faculty and students is being encouraged. Thus, the mobility of scientists in Europe might soon rival that of scientists in the United States, previously one of the strongest qualities of the US research enterprise (see section 4.1).

Individual European countries are spending considerable sums to support interdisciplinary research. For example, the German government has begun experimental programs to increase interdisciplinary training and prepare scientists for nonacademic employment (NSF 1996c, p. 28). In France, megaprojects "grands programmes" with multiyear funding have been financed by the government in fields of scientific priority, and the National Committee for Scientific Research is vigorously supporting collaborative projects in materials science, nanotechnology, and the environment (NSF 1996c, p. 36).

5.3 Employment Prospects for New PhDs

5.3.1 Academic Jobs

Many US universities have experienced severe financial crises during the 1990s for a variety of reasons, such as the general "downsizing" trend in the US economy and lower-than-expected undergraduate enrollments. The consequent unfavorable job market for recent PhDs in science has been discussed in detail in several reports (for example, NSF 1996b, COSEPUP 1995), but no consensus has emerged about ways to solve or even alleviate the unemployment and underemployment of PhDs.

These developments have, not surprisingly, affected mathematics, inasmuch as higher education is the largest US sector that employs mathematics PhDs. (In 1993, jobs in universities and 4-year colleges accounted for 65.2% of all employed US doctoral recipients in the mathematical sciences, followed by 24.8% in private industry and 4.5% in government) (NSF 1996c, table 20). Figure 2 depicts the unemployment rate among new PhDs in mathematics from 1989 to 1995.

The most-prominent reason for the sudden worsening of the job market in 1990 is obvious: an oversupply of new PhDs relative to the availability of tenure-track positions in colleges and universities. The number of PhDs produced by US mathematics departments began to increase in the middle 1980s, rose during the early 1990s, and has shown signs of instability recently, as shown in figure 3 (see section 5.5).

During the same period, the number of academic positions open to new PhDs in mathematics has been shrinking. From 1989 to 1994, the

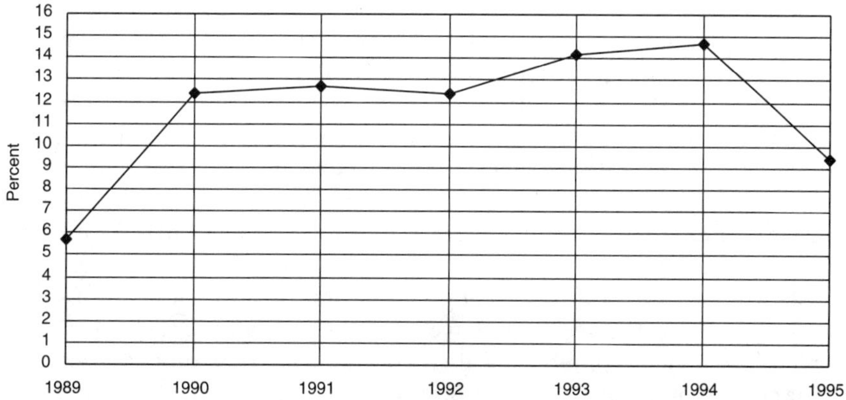

FIGURE 2 Percentage unemployment among new US PhDs in mathematics, autumn of year shown. Source: AMS 1996.

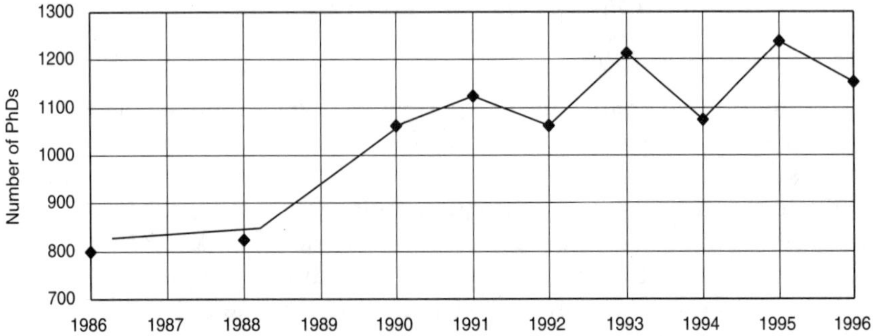

FIGURE 3 Number of PhDs produced by US mathematics departments, spring of year shown. Source: AMS 1997c.

number of positions offered in US mathematics departments to new PhDs fell by 33%; from 1995 to 1996, there was a 6% drop in the number of new PhDs employed by US academic institutions (Davis 1997). During 1994-1995, there were 240 tenure-track positions for new doctoral recipients in US doctorate-granting departments in the mathematical sciences and 184 non-tenure-track positions.

Beyond the diminishing number of academic jobs for new PhDs lies a phenomenon that seems particularly prevalent in US mathematics: a growth in nonpermanent positions. In 1994-95, temporary positions accounted for 50% of the openings for new PhDs in doctorate-granting departments of mathematics. In the autumn of 1996, 64% of the 256 new PhDs who found jobs in academic institutions were in non-tenure-track positions; of those employed in doctorate-granting departments, 84.2% were in non-tenure-track jobs. Overall, the number of full-time US faculty not eligible for tenure rose by 29% from 1991 to 1995.

The existence of an underclass of PhDs who continue to work from year to year at low wages in nonpermanent jobs has led to frustration among recent PhDs (Davis 1997). There has been some recent growth in the number of postdoctoral positions, alleviating unemployment and at the same time providing much further training for fresh PhDs. The law abolishing retirement at a fixed age, which recently began affecting those in academic positions, might further diminish the number of job openings. Most other countries have a fixed retirement age. The pressure on the concept of tenure is likely to increase. Data on the employment situation in other countries are unavailable, but anecdotal information indicates that the problems experienced here in the academic job market are also occurring in other countries.

5.3.2 *Industrial Jobs*

The industrial employment market presents a mixed picture. As shown in figure 4, industrial employment of mathematicians has been increasing. But general trends in industrial research indicate a decrease in spending. Since 1988, industrial spending on research and development in the United States has not increased substantially in constant dollars. In addition, less and less is spent on longer-term research; basic research constituted 6% of industrial expenditures for research and development in 1988 and 2% in 1993 (Council on Competitiveness 1996). Industry's expenditures on basic research declined at an annual average constant-dollar rate of 4.6% from 1991 to 1995 (NSB 1996). In contrast, it is interesting to note, mathematics is expected to grow at both AT&T and Bell Laboratories.

The industrial-research funding picture is more optimistic outside the United States. Many European governments are actively encouraging nondefense industrial research and development; details about these

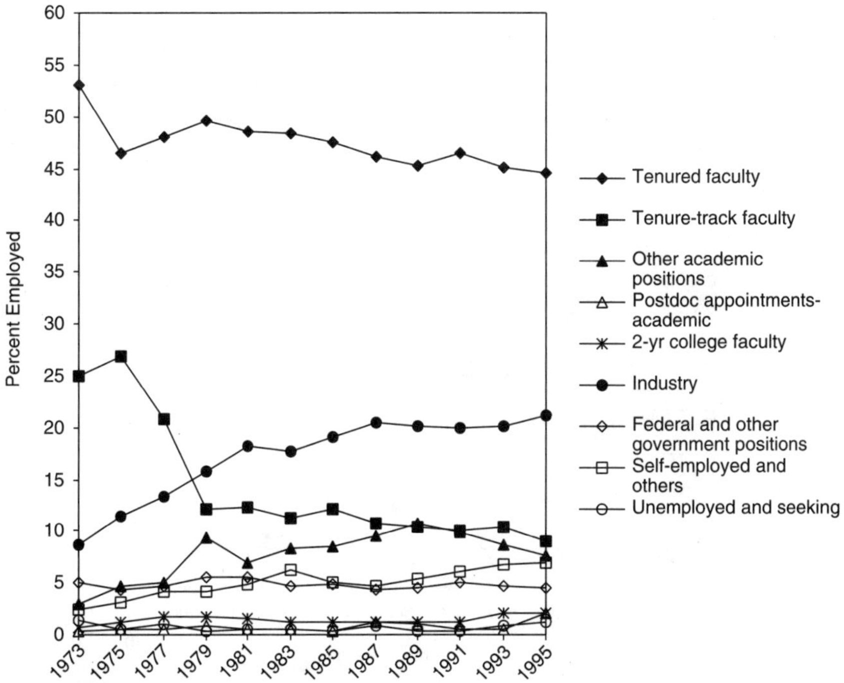

FIGURE 4 Employment status of PhD mathematicians in the US. Source: Analysis conducted by the National Research Council Office of Scientific and Engineering Personnel for this study.

trends can be found in a recent NSF report (1996c). The United States has trailed, for some time, Germany and Japan in civilian research and development as a percentage of GDP; industrial R&D expenditures have been relatively flat in the United States while growing in competitor countries (Council on Competitiveness 1996).

5.4 Foreign Graduate Students

The overall implications of foreign graduate students for US science are discussed in several recent reports, for example, by COSEPUP (1995). Detailed recent data are given by NSB (1996).

During the 1990s, mathematics has been one of the scientific fields most affected by growth in the number and proportion of US PhDs received by non-US students. As figure 5 shows, the number of non-US PhD recipients increased by 78% from 1985 to 1995. Furthermore, in every year since 1990, foreign students have received more than half the PhDs awarded in mathematics in the United States.

That phenomenon occurs elsewhere, and high proportions of foreign students in the sciences are relatively common in other industrialized countries, especially those with former colonial ties. The percentages of foreign natural-science doctoral students in several countries are depicted in figure 6. The large increase shown for Japan is due to Japan's strategy to attract and train foreign students.

A closely related issue is the number of foreign-born PhD recipients who remain permanently in the United States. The panel found no data on how many foreign students receiving mathematics PhDs intend to

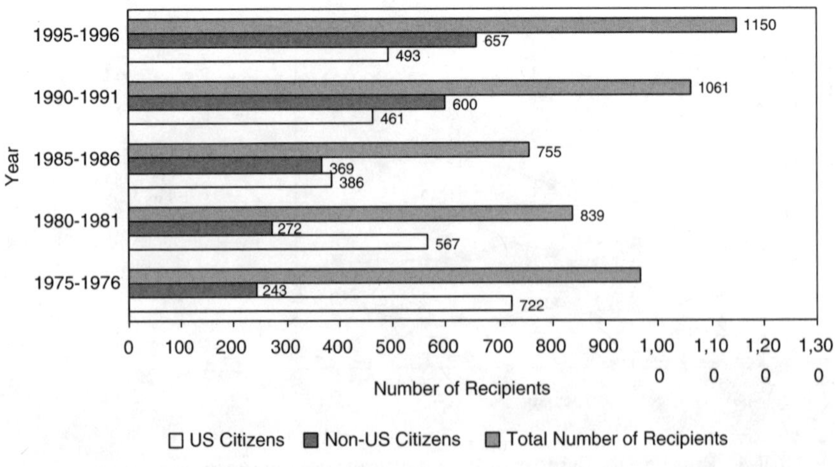

FIGURE 5 Doctoral recipients: total number and US and non-US citizens. Source: AMS 1996.

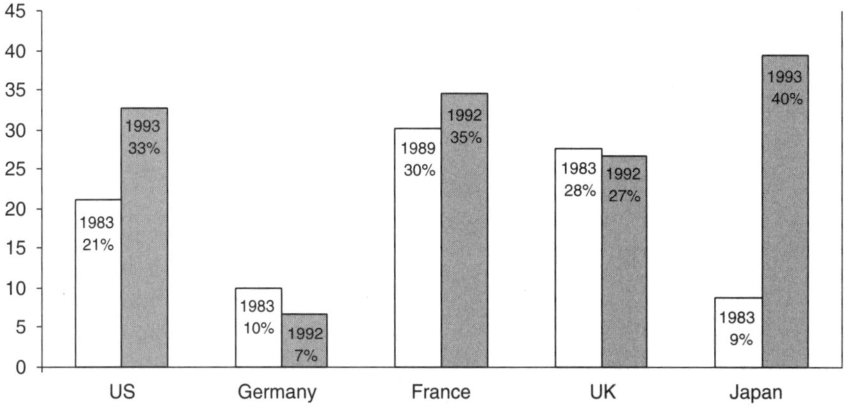

FIGURE 6 Percentage of foreign natural-science doctoral students in various countries. Source: NSB 1996, appendix table 2-33.

remain in the United States after receiving their degrees. However, the overall picture of "stay rates" for foreign students in all science and engineering fields, as shown in figure 7, suggests that such intentions are widespread and confirms the attractiveness of the United States to foreign talent mentioned in section 4.1.

To explore the question further, the panel conducted its own informal survey of 10 highly rated US mathematics departments. Of 397

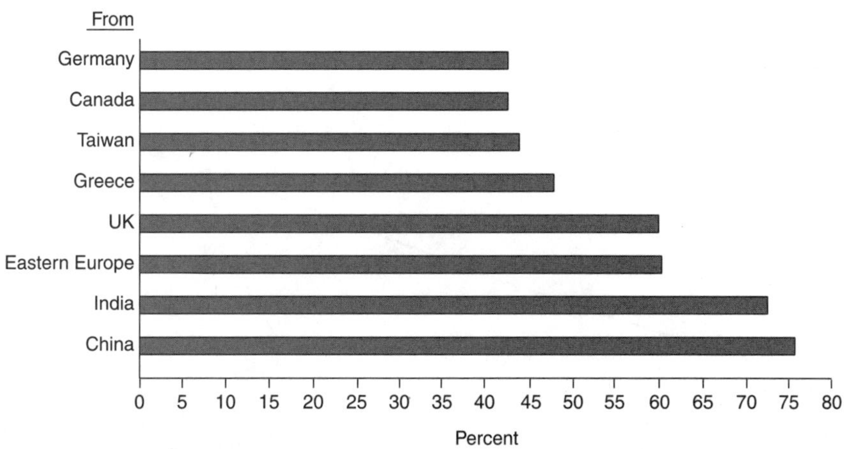

FIGURE 7 Stay rates-percentages of foreign doctoral students who plan to remain in the United States, averaged over 1988-1992. Source: NSB 1996, table 2-15.

tenured faculty, 21% received their undergraduate degree outside the United States; for 107 tenure-track faculty, this statistic was 58%. Thus, the number of faculty in US mathematics departments with undergraduate degrees from outside the United States can be expected to increase.

Stay rates in other countries were found only for France, where 56% of non-French people who received mathematics PhDs in 1992 remained in France (NSF 1996c).

5.5 Graduate Education

As discussed in section 5.3.1, the number of PhDs produced by US universities grew substantially from the middle 1980s through the 1990s. However, the trend has recently changed as doctorate-granting institutions have begun to reduce the size of their graduate programs. In particular, in the autumn of 1996, the projected size of the new class of PhD students in mathematics at US universities was 2,384 compared with 2,546 in the autumn of 1994. Figure 8 shows the total population of full-time doctoral students in mathematics for 1980, 1985, and the 1990s. Since a high in 1992, the number of full-time PhD students in mathematics has steadily decreased.

An online NSF data brief of February 1997 (NSF 1997a) reveals that, among all US doctoral students in the sciences, the largest percentage from 1994 to 1995 occurred in the mathematical sciences and physics, each of which experienced a 6% reduction.

The decreases in applications by both US and non-US students are dramatic, although it is unknown whether they signal the beginning of a trend. Interest in obtaining a PhD in mathematics appears to have been affected by the employment prospects described in section 5.3 for both US and non-US students. A very recent set of data (AMS 1997a, b)

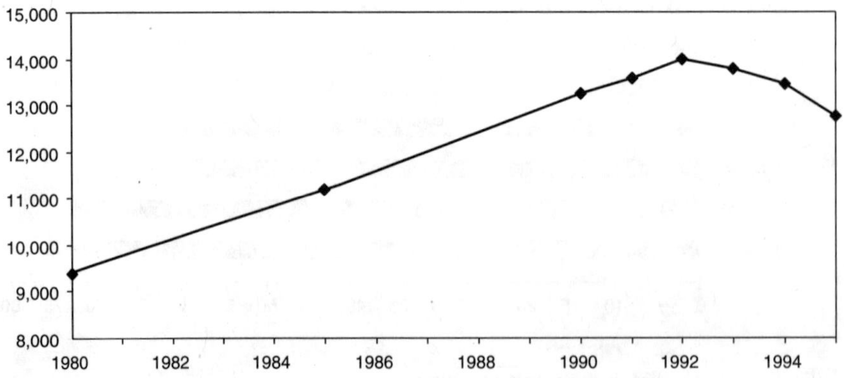

FIGURE 8 Total full-time PhD students in mathematical sciences. Source: NSF 1995.

TABLE 1 Decrease in applications to PhD programs in mathematics, 1994 to 1996

	Top-Ranked Departments	All Departments
Total pool, 1996	7,366	16,516
Total pool, 1994	10,320	23,545
Percentage Decrease, 1994 to 1996	29%	30%
US pool, 1996	3,108	6,291
US pool, 1994	4,769	9,270
Percentage decrease, 1994 to 1996	35%	32%
International pool, 1996	4,295	10,387
International pool, 1994	5,498	14,537
Percentage decrease, 1994 to 1996	22%	29%

Source: AMS 1997a, pp. 213-216.
Note: The total pool may not equal the sum of the US pool and the international pool. Since some departments were unable to provide numbers of applications broken out by citizenship or visa status, the projections may be based on slightly different sets of respondents. Top ranked departments are those offering the PhD and which have high "scholarly quality of program faculty" as reported in the 1995 National Research Council report *Research-Doctorate Programs in the United States: Continuity and Change* (NRC 1995d). There are 48 top-ranked departments.

collected in mid-1996 shows that there has been a uniform drop in applications to mathematics graduate schools from 1994. Table 1 shows data on the 48 top-ranked mathematics departments and on all doctoral programs in mathematics. Other reasons for the decline might be competition from computer science, biologic science, and medicine and poor preparation in high school and college.

Another issue is the degree to which women and members of minority groups are pursuing graduate degrees in mathematics. From 1983 to 1993, the percentage of new PhDs who were women grew from 16.1% to 23%; this is slightly greater than the percentage for all the physical sciences and computer science. The percentage of minority-group members receiving mathematics PhDs is much smaller. For example, only 8 of some 583 mathematics PhDs awarded to Americans went to blacks in 1993, and this number has remained roughly constant over the last decade. The situation for Hispanic Americans is a bit different: 16 received degrees (NSF 1996b).

No data were found on the size of graduate mathematics programs in other countries.

5.6 Support

In section 4.4. we stated that an important underpinning for US success in mathematical research has been sustained support and

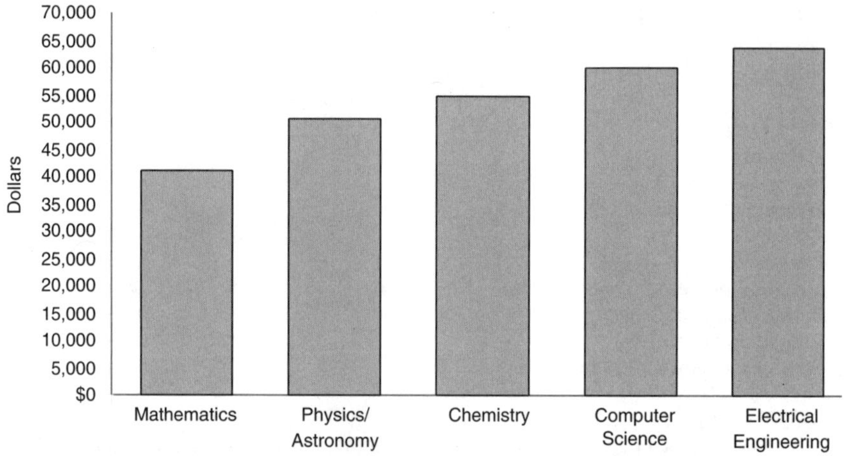

FIGURE 9 Median salaries in 1993 of US PhDs who received their degrees in 1985-1990, by field. Source: NSF 1996a, appendix table 5-27.

funding. Before choosing to obtain a PhD in mathematics, the most-talented people are likely to consider not simply their expected salary, but also their likelihood of receiving support for the time and resources needed to carry out their research.

Figure 9 compares the 1993 median salaries of US PhDs who received their degrees in 1985-1990 in mathematics, computer science, chemistry, physics/astronomy, and electrical engineering. One might reasonably conclude that mathematics PhDs have less-favorable salary prospects than other science PhDs. We have no comparable data for other countries.

It is difficult to make international comparisons with respect to salaries and federal support because university researchers in other countries do not typically receive summer salary support from individual government grants. In the UK and Canada, for example, academic salaries are paid entirely by universities.

As to federal research support, figure 10 shows that a lower percentage of academic mathematicians received US federal support in 1993 than any other category of doctoral scientists except social scientists.

Finally, the mathematical sciences have not fared well, compared with other sciences, in overall federal support in recent years (see figure 11). For example, in 1994-1995, overall federal support for academic research and development grew by 5%, but support for the mathematical sciences dropped relative to that for other sciences. Mathematics had the lowest rate of growth (1%) in federal funding for research and was

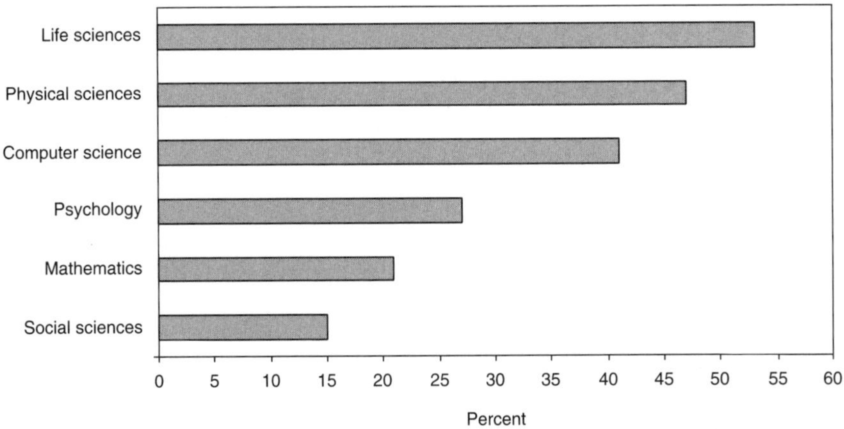

FIGURE 10 Percentages of academic scientists with federal support, 1993. Source: NSB 1996, appendix table 5-27.

the only science whose support grew at a rate lower than that of inflation, which was 1.8% (NSF 1997b).

The details of the picture vary by agency. On the basis of current dollars in the actual FY1996 and estimated FY1997 budgets, the Divi-

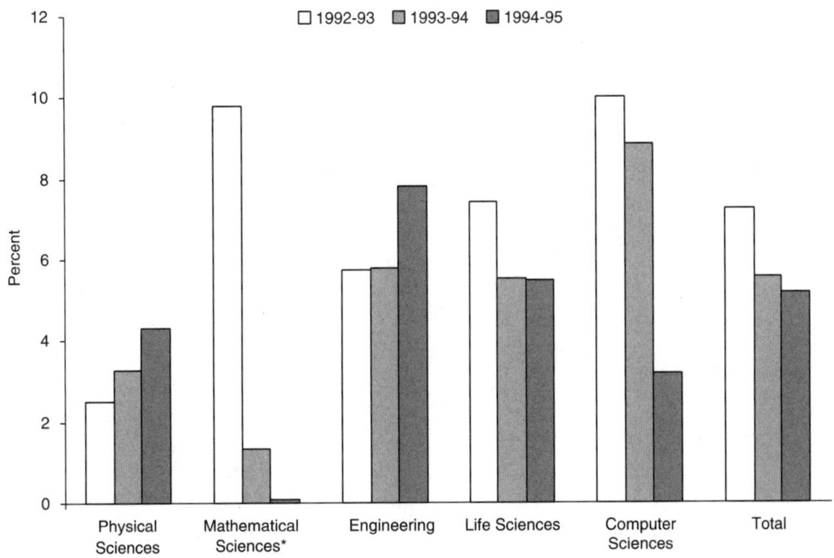

FIGURE 11 Percentage increase in federal R&D expenditures at universities and colleges, by field. Source: NSF 1997b, table 1 and discussions with NSF staff.

sion of Mathematical Sciences at the National Science Foundation experienced growth of 7.1%. Overall Department of Defense spending on mathematical sciences decreased by 12.3% and overall Department of Energy spending on mathematical sciences remained flat.

6

LIKELY FUTURE RELATIVE POSITION OF US MATHEMATICS

The current trends described in section 5 are obviously mixed. This section summarizes our best estimate about the future relative position of US mathematical research.[1]

6.1 Intellectual Quality

As already stressed in section 5.1., the field is full of new results, new methods, new points of view, and new problems. Because the United States is preeminent, mathematics in the United States is likely in the near term to retain its dominant position in the world. In the long term, however, some of this momentum might be lost, depending on how we rise to new challenges and potentially damaging developments described below.

6.2 Interdisciplinary Research

Notable successes in interdisciplinary research (see sections 3.2.1. and 3.2.2.) have made this aspect of US mathematical research of preeminent importance today—although not uniformly throughout the United States. As recognition of the importance of mathematics in interdisciplinary research grows, opportunities will expand for collaborations that enrich other sciences and mathematics. The panel believes that the future relative position of the United States in interdisciplinary mathematical research depends in large part on the effectiveness with which these opportunities are realized. As observed in sections 5.2 and

[1]Many of these issues are discussed in a National Research Council report (NRC 1997).

5.3.2, governments, universities, and industry in other countries are actively supporting mechanisms that encourage interdisciplinary research. The United States must pay serious attention to this issue.

6.3 US Graduate Education in Mathematics

The panel is especially concerned about the potential erosion of the US research base because of a decrease in the number of graduate students at leading universities. The trends discussed in section 5.4 imply that the future position of US mathematics is likely to depend increasingly on graduate students and postdoctoral fellows from other countries; this makes our preeminence precarious if jobs in their countries of origin become more attractive to foreign students or if changes in immigration laws close the United States doors to non-native mathematicians. In addition, there is a dearth of minority-group members in mathematics. The panel believes that the United States must cultivate its own mathematical talent to retain its leading stature in mathematical research. A key factor in this cultivation is the quality of mathematics education in K-12 and college.

6.4 Support for Mathematical Research

The most important safeguard of US preeminence in mathematical research—and in all the sciences—is the flourishing of both private and state research universities. Some of the stresses faced today by US research universities are described by the Council on Competitiveness as "facing a funding squeeze and growing, often contradictory, demands" (Council on Competitiveness 1996, p. 21). The research universities respond to this squeeze in part by reducing staff size. This situation affects mathematics because research universities provide a stable base, both financial and professional (see section 4.4). The current trend toward hiring temporary faculty discussed in section 5.3.1 is a prime indication that US universities might provide much less of that support in the future.

Today the research universities are the major instruments in the United States for research and development that fuel high technologies, an extremely important part of the US economy. Mathematics has prospered in part because it plays an important role in this research. But the research enterprise is at risk if the support for research universities continues to decline.

7

REFERENCES

ABIE (Australian Bureau of Industry Economics). *Australian Science: Performance from Published Papers*. Report 96/3. Canberra, Australia: Australian Government Publishing Service, 1996.

AMS (American Mathematical Society). "AMS Task Force on Employment, Report to the Mathematical Community Part I: Academic Hiring Survey, 1991-1992", Notices of the AMS, April 1992.

AMS (American Mathematical Society). "1996 Annual AMS-IMS-MAA Survey (First Report)", Notices of the AMS, December 1996.

AMS (American Mathematical Society). "Recent Trends in Graduate Admissions in Mathematics Departments", Notices of the AMS, February 1997a.

AMS (American Mathematical Society). "A Survey of Four-Year and University Mathematics in Fall 1995: A Hiatus in Both Enrollment and Faculty Increases", Notices of the AMS, September 1997b.

AMS (American Mathematical Society). "1996 Annual AMS-IMS-MAA Survey (Second Report)", Notices of the AMS, September 1997c.

Askey, Richard. "The Japanese University Entrance Examinations in Mathematics", FOCUS Newsletter, Mathematics Association of America, vol. 13, no. 1, 1993.

Chen, L.Q., B. Fultz, J.W. Cahn, J.R. Manning, J.E. Morral, J. Simmons, eds. *Mathematics of Microstructure Evolution.* The Minerals, Metals, and Materials Society, Warrendale, PA, and Society for Industrial and Applied Mathematics, Philadelphia, PA, 1996.

COSEPUP (Committee on Science, Engineering, and Public Policy). *Science, Technology, and the Federal Government: National Goals for a New Era*. Washington, D.C.: National Academy Press, 1993.

COSEPUP (Committee on Science, Engineering, and Public Policy). *Reshaping the Graduate Education of Scientists and Engineers.* Washington, D.C.: National Academy Press, 1995.

COSEPUP (Committee on Science, Engineering, and Public Policy). *An Assessment of the National Science Foundation's Science and Technology Program.* Washington, D.C.: National Academy Press, 1996.

Council on Competitiveness. *Endless Frontier, Limited Resources.* Washington, D.C.: Council on Competitiveness, 1996.

Davis, Geoff. "Mathematicians and the Market", March 30, 1997. http://www.cs.dartmouth.edu/~gdavis/policy/papers.html.

ED (U.S. Department of Education). *Pursuing Excellence*, NCES 97-198. National Center for Education Statistics. Washington, D.C.: U.S. Government Printing Office, 1996.

NAE (National Academy of Engineering). *Frontiers of Engineering*. Washington, D.C.: National Academy Press, 1996.

NAPA (National Academy of Public Administration). *National Science Foundation's Science and Technology Centers: Building an Interdisciplinary Research Paradigm*. Washington D.C.: National Academy of Public Administration, 1995.

NRC (National Research Council). *Mathematical Sciences, Technology, and Economic Competitiveness.* Washington, D.C.: National Academy Press, 1991.

NRC (National Research Council). *Mathematical Research in Materials Science*. Washington, D.C.: National Academy Press, 1993.

NRC (National Research Council). *Mathematical Challenges from Theoretical/Computational Chemistry*. Washington, D.C.: National Academy Press, 1995a.

NRC (National Research Council). *Calculating the Secrets of Life: Applications of the Mathematical Sciences to Molecular Biology.* Washington, D.C.: National Academy Press, 1995b.

NRC (National Research Council). *Allocating Federal Funds for Science and Technology.* Washington, D.C.: National Academy Press, 1995c.

NRC (National Research Council). *Research-Doctorate Programs in the United States: Continuity and Change*. Washington, D.C.: National Academy Press, 1995d.

NRC (National Research Council). *Cryptography's Role in Securing the Information Society.* Washington, D.C.: National Academy Press, 1996a.

NRC (National Research Council). *Mathematics and Physics of Emerging Biomedical Imaging.* Washington, D.C.: National Academy Press, 1996b.

NRC (National Research Council). *Massive Data Sets: Proceedings of a Workshop*. Washington, D.C.: National Academy Press, 1996c.

NRC (National Research Council). *Preserving Strength While Meeting Challenges: Summary Report of a Workshop on Actions for the Mathematical Sciences*. Washington, D.C. National Academy Press, 1997.

NSB (National Science Board). *Science and Engineering Indicators - 1996*. Arlington, VA: National Science Foundation, 1996.

NSF (National Science Foundation). *Characteristics of Doctoral Scientists and Engineers in the United States, 1993.* NSF 96-302. Arlington, VA: National Science Foundation, 1996a.

NSF (National Science Foundation). *Women, Minorities, and Persons with Disabilities in Science and Engineering: 1996*. NSF 96-311. Arlington VA: National Science Foundation, 1996b.

NSF (National Science Foundation). *Human Resources for Science & Technology: The European Region*. NSF 96-316. Arlington, VA: National Science Foundation, 1996c.

NSF (National Science Foundation). Data Brief. 97-303. Arlington, VA: National Science Foundation, February 7, 1997a.

NSF (National Science Foundation). Data Brief. 97-304. Arlington, VA: National Science Foundation, March 7, 1997b.

NSF (National Science Foundation). Knowledge and Distributed Intelligence. See World Wide Web documents at http://www.nsf.gov.

NSF (National Science Foundation). 1995 Graduate Student Survey. Available at http://www.nsf.gov.

SIAM (Society for Industrial and Applied Mathematics). *Mathematics in Industry.* Philadelphia: SIAM, 1995.

SIAM (Society for Industrial and Applied Mathematics). "Mathematics meets film animation in Kansas City community lecture", SIAM News, vol. 29, no. 7, 1996.

United Kingdom Office of Science and Technology. *The Quality of the UK Science Base*. London, UK: Department of Trade and Industry, March 1997.

APPENDIX A

PANEL AND STAFF BIOGRAPHICAL INFORMATION

Panel

Peter D. Lax [NAS*] *(Chair)* is professor of mathematics and director of the Courant Mathematics and Computer Lab at New York University. Before taking his current positions, he was director of the Courant Institute of Mathematical Sciences (1972-1980) and director of the AEC Computing and Applied Mathematics Center (1964-1972), both at NYU. He is a National Medal of Science awardee, former president of the American Mathematical Society (AMS), and a former National Science Board member. He has extensive experience as a lecturer overseas, particularly in Germany, France, and the United Kingdom as well as being a foreign member of the French, Chinese, Hungarian, and Russian academies of sciences. He is a member of the American Philosophical Society. Dr. Lax has received many honorary degrees, the Wolf Prize (1987), the National Academy of Sciences Prize in Applied Mathematics (1983), the Norbert Wiener Prize of AMS and the Society for Industrial and Applied Mathematics (1975), the Chauvenet Prize of the Mathematical Association of America (1974), and the Steele Prize of AMS (1992).

Michael F. Atiyah [NAS-F] is director of the Isaac Newton Institute for Mathematical Sciences in Cambridge, England, and president of the Royal Society (1990-1995). He has recently retired as master of Trinity College, Cambridge. Before this, he was a Royal Society research professor and fellow of St. Catherine's College, Oxford (1973-1990). He has served as professor of mathematics at the Institute for Advanced

Study in the United States (1969-1972) and Savilian Professor of Geometry at University of Oxford (1963-1969). He attended Victoria College in Cairo and received his BA (1952) and PhD (1955) from the University of Cambridge. Sir Michael is a member of many academies of sciences. He holds many honorary doctorates and several honorary fellowships. Sir Michael has been a member of the Executive Committee of the International Mathematical Union (1966-1974), president of the London Mathematical Society (1975-1977), president of the Mathematical Association (1981-1982), member of the Science and Engineering Research Council (1984-1989), member of the Council of the Royal Society (1973-1974), vice-president of the Royal Society (1984-1985), and chairman of the European Mathematical Council (1978-1990). He received the Fields Medal in 1966.

Spencer J. Bloch [NAS] is professor of mathematics at the University of Chicago. He was an instructor, lecturer, and assistant professor at Princeton (1971-1974) and an associate professor at the University of Michigan (1974-1976). Dr. Bloch has done pioneering work in the application of higher algebraic K theory to algebraic geometry, particularly in problems related to algebraic cycles, and is regarded as the world's leader in this field. His work has firmly established higher K theory as a fundamental tool in algebraic geometry. Dr. Bloch received his BA at Harvard (1966) and his PhD from Columbia (1971).

Joseph B. Keller [NAS] is professor of mathematics and mechanical engineering (emeritus) at Stanford University. Before this, he was at Courant Institute of Mathematical Sciences, New York University (1948-1979), where he received his PhD in mathematics (1948). He is a foreign member of the Royal Society and was honorary professor of mathematical sciences at the University of Cambridge. He is a recipient of the Wolf Prize (1997), the Frederick E. Nemmers Prize (1996), the National Academy of Sciences Award in Applied Mathematics and Numerical Analysis (1995), the National Medal of Science (1988), the Timoshenko Medal (1984), the Eringen Medal (1981), and the von Karman Prize (1979). He was von Neumann Lecturer (1983) and Gibbs Lecturer of the American Mathematical Society (1977).

Jacques-Louis Lions [NAS-F] is professor at the College de France in Paris and president of the French Academy of Sciences. He has contributed fundamental research in nonlinear partial differential equations, including homogenization and control. He has trained an entire generation of modern applied mathematicians in France. He has pioneered industrial and applied mathematics cooperation in France as cofounder of Institut National de Recherche en Informatique et en Automatique (INRIA) and as president of the French Space Agency. In addition, he is a former president and secretary of the International Mathematical Union.

Yuri I. Manin is a member (since 1993) and director (since 1995) of the Max Planck Institute for Mathematics. He is also leading researcher of the Steklov Mathematical Institute of the Russian Academy of Sciences (since 1960, now in absentia.) In 1965-1992, he was professor of algebra at Moscow University and held various visiting professorships, in particular at Harvard University, Columbia University, and the Massachusetts Institute of Technology. He is a member of the Academy of Sciences, Russia, the Royal Society of Sciences of the Netherlands, the Academia Europaea, the Max-Planck-Gesellschaft, the Gottingen Academy of Sciences Class of Physics and Mathematics, and the Pontificia Academia Scientiarum. He won the Lenin Prize for work in algebraic geometry (1967), the international Frederic Esser Nemmers Prize in Mathematics of Northwestern University (1994) and the Brouwer Golden Medal of the Royal Society and Mathematical Society of the Netherlands for work in number theory (1987).

Rudolph A. Marcus [NAS] is A.A. Noyes Professor of Chemistry at the California Institute of Technology. He is also an honorary professor at Fudan University in Shanghai, China, and at the Institute of Chemistry in the Chinese Academy of Sciences in Beijing. Dr. Marcus holds an honorary fellowship at University College of the University of Oxford and was Linnett Visiting Professor of Chemistry at the University of Cambridge. He previously held positions at the University of North Carolina, the Polytechnic Institute of Brooklyn, the Courant Institute of Mathematical Sciences of New York University, the University of Illinois, the University of Oxford, and the California Institute of Technology. Dr. Marcus received his BSc (1943) and PhD (1946) in chemistry from McGill University in Montreal, Canada. He is a fellow of the National Academy of Sciences, the American Academy of Arts and Sciences, the Royal Society of Chemistry, and the Royal Society of Canada, and he is a member of the Royal Society, the International Academy of Quantum Molecular Science, the American Philosophical Society, the American Association for the Advancement of Science, the American Chemical Society, and the American Physical Society. Dr. Marcus received the Nobel prize in chemistry in 1992.

Gary C. McDonald is head of the Operations Research Department at the General Motors Research and Development Center. Before this, he was head of the mathematics department at the center (1983-1992). He is also adjunct professor of mathematics at Oakland University. Dr. McDonald started as associate senior research mathematician at General Motors in 1969 and has held the positions of senior research mathematician (1972-1976) and assistant department head (1976-1983). He received his BA (1964) from St. Mary's College and his MS (1966) and PhD (1969) from Purdue University. He is a fellow of the Institute of Mathematical Statistics, the American Statistical Association, and the American Association for the Advancement of Science.

Cathleen S. Morawetz [NAS] is professor emeritus at the Courant Institute of Mathematical Sciences of New York University. She has been with the institute since starting as a research associate (1952). She held positions as assistant professor (1957-1960), associate professor (1960-1965), professor (1965-present), and chairman (1981-84) Department of Mathematics, associate director (1978-1981), deputy director (1981-1984), and director (1984-1988). Dr. Morawetz was a trustee of the Alfred P. Sloan Foundation (1980-1984) and a member of the National Research Council's Board on Mathematical Sciences. She was president of the American Mathematical Society during 1995-1996 and is a member of the Mathematical Association of America, and the Society for Industrial and Applied Mathematics and a fellow of the American Association for the Advancement of Science. She received her BA (1945) from the University of Toronto, her MS (1946) from the Massachusetts Institute of Technology, and her PhD (1951) from New York University.

Peter Sarnak is chairman of the Department of Mathematics at Princeton University. Before this, he was the H. Fine Professor (1995-1996). He has also been professor at Stanford University (1987-1991), the Sherman Fairchild Distinguished Scholar at the California Institute of Technology (1989), a fellow at the Institute of Advanced Studies at Hebrew University (1987-1988), and assistant and associate professor at the Courant Institute of Mathematical Sciences of New York University (1980-1983). Dr. Sarnak received his BSc (1974) from the University of Witwatersrand in South Africa and his PhD (1980) from Stanford University. He is a member of the American Academy of Arts and Sciences and was a Sloan fellow (1983-1985) and presidential young investigator (1985-1990).

I.M. Singer [NAS] is Institute Professor at the Massachusetts Institute of Technology (MIT). His research has been in the fields of index theory/manifold invariants/elliptic analysis, differential geometry, functional analysis, and operator theory. He has been teaching calculus intermittently between 1949 and 1997. He received the AMS Bocher Memorial Prize (1969), the National Medal of Science (1983), the Wigner Medal (1988), and the AMS Award for Public Service (1993). He was chairman of the National Academy of Sciences Committee on Science and Public Policy (1973-1978) and was with the White House Science Council from

1982 to 1988. Dr. Singer received his BS (1944) from the University of Michigan and his MS (1948) and PhD (1950) in mathematics from the University of Chicago. He is a member of the Council of the National Academy of Sciences, the Governing Board of the National Research Council, the American Academy of Arts and Sciences, the American Mathematical Society (vice president, 1970-1972), the American Physical Society, and the American Philosophical Society.

Margaret H. Wright [NAE] is a Distinguished Member of Technical Staff at Bell Laboratories, Lucent Technologies. Before joining Bell Laboratories she worked in the Department of Operations Research at Stanford University (1976-1988). Dr. Wright holds a BS in mathematics, and MS and PhD degrees in computer science, from Stanford University. She served as president of the Society for Industrial and Applied Mathematics during 1995-1996. Her research involves theory and algorithms for optimization and linear algebra, scientific computing, and solution of real-world optimization problems.

Staff

Deborah D. Stine is the study director and associate director of the Committee on Science, Engineering, and Public Policy (COSEPUP). She has been working on various projects throughout the National Academy of Sciences complex since 1989. She received a National Research Council group award for her first study for COSEPUP on policy implications of greenhouse warming and a Commission on Life Sciences staff citation for her work in risk assessment and management. Other studies have addressed graduate education, responsible conduct of research, careers in science and engineering, environmental remediation, the national biological survey, and corporate environmental stewardship. She holds a bachelor's degree in mechanical and environmental engineering from the University of California, Irvine; a master's degree in business administration; and a PhD in public administration, specializing in policy analysis, from the American University. Before coming to the Academy, she was a mathematician for the Air Force, an air-pollution engineer for the state of Texas, and an air-issues manager for the Chemical Manufacturers Association.

John R. Tucker has been director of the Board on Mathematical Sciences (BMS) since 1994. He earned degrees in mathematics at Washington College (BA) and George Washington University (Mphil and PhD). He has worked as a researcher with Chi Associates, Inc., and an assistant professor of mathematics at Virginia Commonwealth University and Mary Washington College. He joined the National Research Council in 1989 as program officer for the BMS and advanced to senior program officer in 1993. His interests include nonlinear dynamics, order and disorder, mixing processes, and mathematical developments in biology and medicine.

Lawrence E. McCray is executive director of the Committee on Science, Engineering, and Public Policy (COSEPUP). He held positions in the Environmental Protection Agency, the US Regulatory Council, and the Office of Management and Budget before coming to the National Academy of Sciences in 1981. He has directed Academy studies in carcinogenic risk assessment, export controls, nuclear winter, and federal science budgeting. A Fulbright scholar in 1968, he received the Schattschneider award in 1972 from the American Political Science Association for the best dissertation in American government and politics. In 1987, he received the National Research Council staff award. He joined COSEPUP in 1988 as executive director and since 1994 has served concurrently as the director of the NRC Policy Division.

Patrick P. Sevcik is research associate with the Committee on Science, Engineering, and Public Policy (COSEPUP). He works on a variety of projects for COSEPUP, the Policy Division (PD), and the PD Office of Special Projects, assisting Deborah Stine and Lawrence McCray. Before coming to the National Research Council in 1993, he was an assistant program officer with the International Republican Institute from 1990 to 1993, working on democracy development, primarily in central and eastern Europe. He has held positions at the White House in the Office of Political Affairs (1989-1990) and on Capitol Hill (1987-1988) in the office of Representative John DioGuardi (R-NY). During that time, he also held concurrent positions in several Slovak-American organizations. He holds a BA in international affairs, with an emphasis on Soviet and Eastern European studies, from the George Washington University. He has also studied Russian language and culture at the Leningrad Polytechnic Institute in Leningrad.

---

*	NAS	Member of the National Academy of Sciences
	NAS-F	Foreign member of the National Academy of Sciences
	NAE	Member of the National Academy of Engineering

APPENDIX B

STATISTICAL DATA ON THE FIELD OF MATHEMATICS

This appendix is a collection of some of the data that various members of the panel reviewed before developing conclusions. It provides the available data on education, employment, funding, and papers and citations. Most of the information is available only for the United States, but non-US data, when available, are included.

Education

Figure B-1 shows how the number of institutions in the United States awarding PhDs in mathematics has grown since 1920. Figure B-2 provides the number of PhDs that these institutions awarded during the same period. The drastic increase in PhDs in the 1960s was probably due to the draft exemption during the Vietnam War. The big increase in degrees granted in the 1980s probably occurred when computer science came into vogue.

Figure B-3 shows how long it took students to attain their degrees and provides the age at which they received their doctorate. Figure B-4 shows how many of those students were foreign citizens, and table 1 in section 5.5 shows the decrease in applications to US PhD programs in mathematics by US and non-US citizens.

Figure B-5 compares the number of first degrees (equivalent to a BS in the United States) in mathematics and computer science in the United States and western Europe. The data were available only for mathematics and computer science combined, and computer science grew rapidly during the period covered, especially in western Europe.

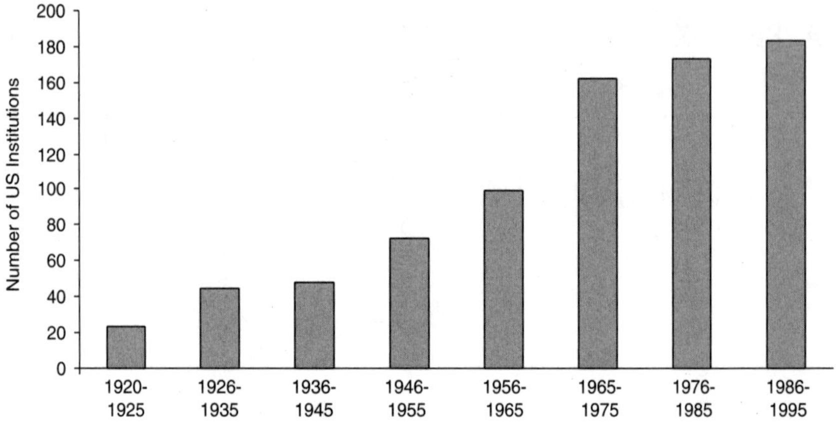

FIGURE B-1 Number of US institutions awarding PhDs in mathematics, 1920-1995. Source: Analysis conducted by the National Research Council's Office of Scientific and Engineering Personnel for this study.

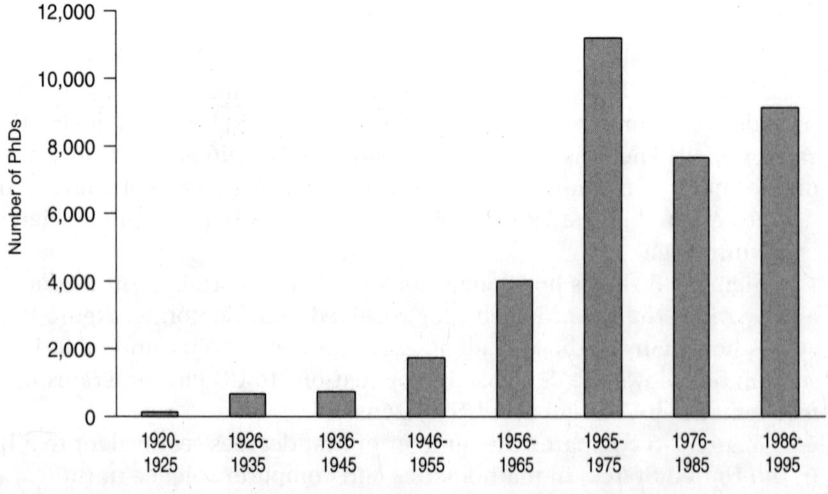

FIGURE B-2 Number of PhDs awarded in mathematics in the United States, 1920-1995. Source: Analysis conducted by the National Research Council's Office of Scientific and Engineering Personnel for this study.

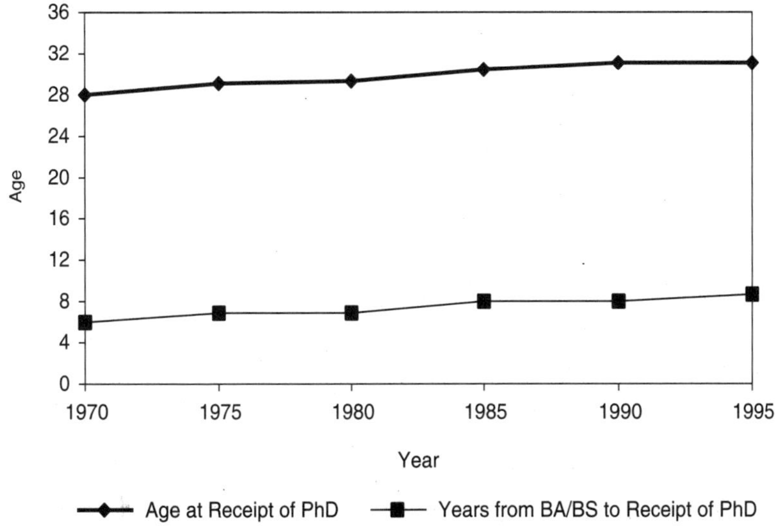

FIGURE B-3 Median time to PhD and age at receipt of PhD in mathematics in the United States. Source: COSEPUP 1995.

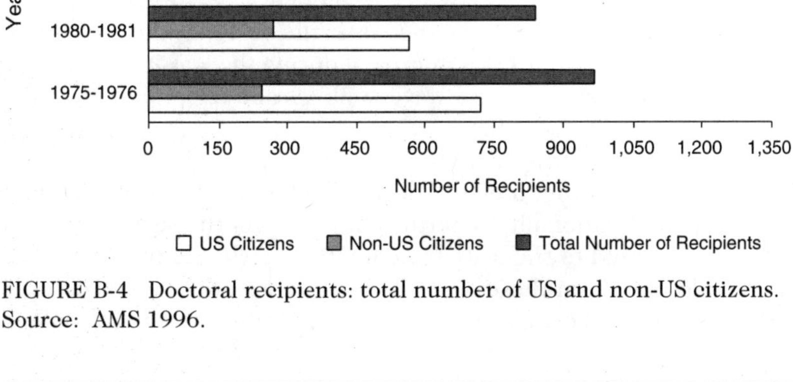

FIGURE B-4 Doctoral recipients: total number of US and non-US citizens. Source: AMS 1996.

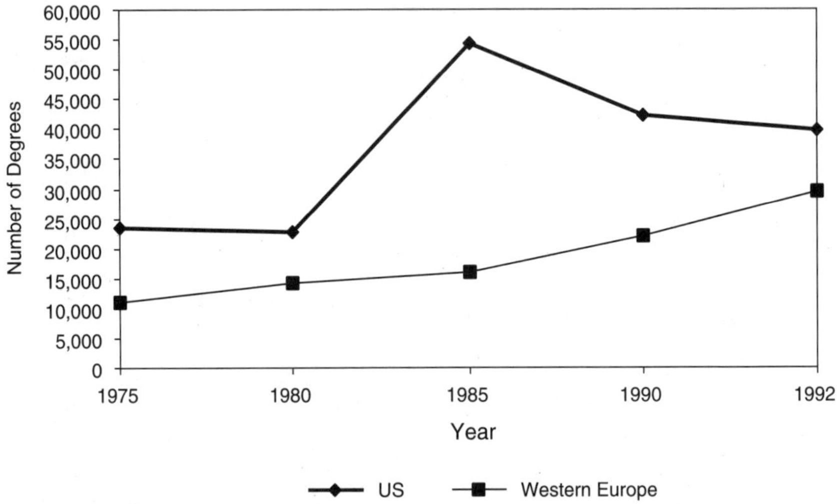

FIGURE B-5 Number of first degrees in mathematics and computer science. Source: NSF 1996c, p. 34.

Figure B-6 shows the number of doctoral degrees awarded in natural sciences in Asia, Europe, and the United States in 1992. Mathematics cannot be separated out from these data.

EMPLOYMENT

Figure B-7 shows the number of PhD mathematicians employed in the United States from 1973 to 1991. Where they are employed is shown in table B-1, and the type of work they are doing is shown in table B-2.

The data are from the Survey of Doctorate Recipients (SDR). The SDR is a biennial longitudinal survey, dating back to 1973, of research doctorates working in the United States. The survey questionnaire is sent in the spring to a sample of about 50,000. These people are asked a series of demographic and employment-characteristics questions. The response rate for the survey has varied over the years; in the late 1980s it was about 60%. That has been improved during the last 2 survey cycles through the use of second-wave mailings and telephone interviews; in 1995, it was about 85%.

The sample is stratified across 3 variables: field of degree, sex, and a combination variable that includes degree field, sex, handicap status, ethnic group, and nationality of birth. The results of the survey are statistically analyzed to translate the data into weighted numbers for the entire population. From these data, the doctorate workforce in science and engineering can be analyzed across different dimensions by looking

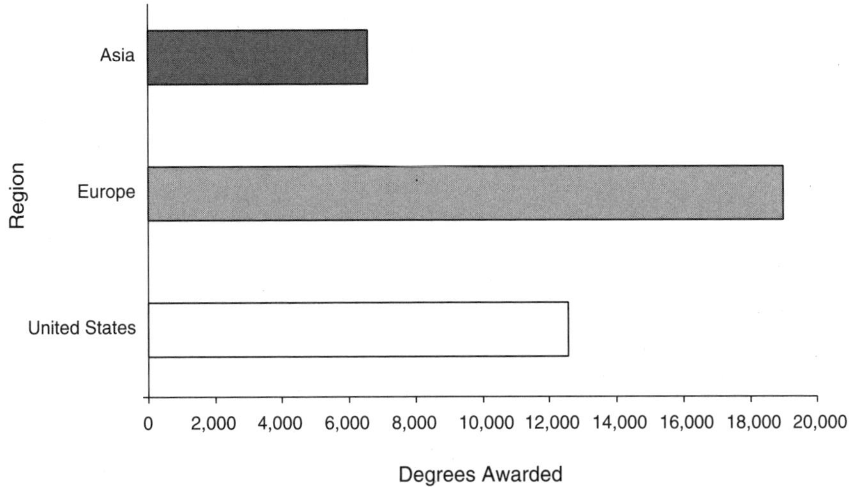

FIGURE B-6 Doctoral degrees in natural-sciences, 1992. Source: NSF 1996c, p. 8.

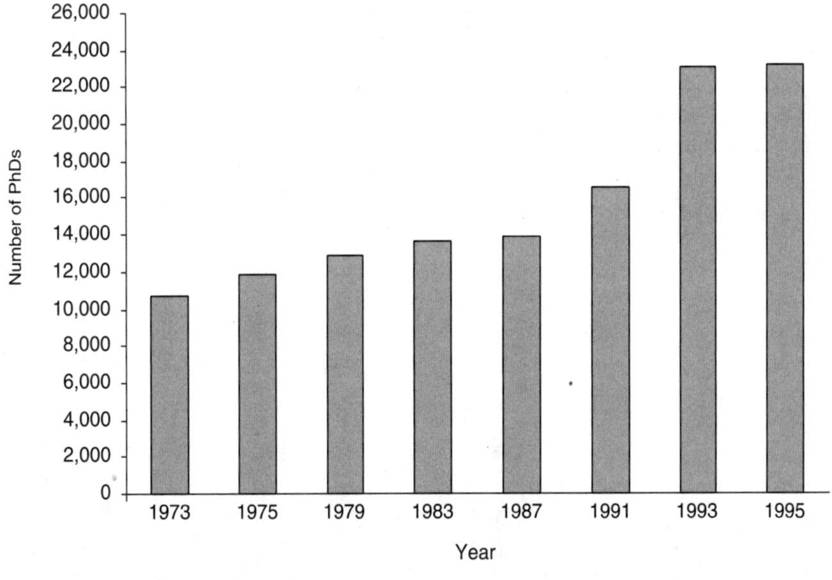

FIGURE B-7 Number of PhD mathematicians employed in the United States. Source: COSEPUP 1995, p. 153.

TABLE B-1 Employment Status of PhD Mathematicians in the United States

	1973	1975	1977	1979	1981	1983	1985	1987	1989	1991	1993	1995
Tenured and tenure-track faculty	9,238	10,332	10,690	10,341	10,936	11,233	11,887	11,627	12,057	12,209	12,779	12,437
Tenured faculty	6,280	6,546	7,458	8,303	8,715	9,124	9,469	9,438	9,801	10,039	10,386	10,338
Tenure-track faculty	2,958	3,786	3,232	2,038	2,221	2,109	2,418	2,189	2,256	2,170	2,393	2,099
Other academic positions	343	666	779	1,582	1,266	1,592	1,693	1,950	2,313	2,133	1,994	1,788
Postdoctoral appointments—academic	41	73	93	146	102	102	83	264	210	99	105	484
2-yr college faculty	87	183	284	290	275	233	233	245	252	257	454	466
Industry	1,027	1,622	2,064	2,635	3,274	3,350	3,795	4,179	4,355	4,295	4,658	4,886
Federal and other government positions	586	601	719	918	988	887	958	871	965	1,076	1,083	1,045
Self-employed and others	280	448	634	686	855	1,195	989	940	1,177	1,292	1,550	1,597
Postdoctoral appointments—other	31	35	31	30	87	33	47	25	45	19	50	27
Unemployed and seeking employment	161	79	168	66	102	105	86	169	91	62	211	301
Elementary- and high school teachers	22	35	49	52	102	172	130	196	172	192	123	155
TOTAL	11,816	14,074	15,511	16,746	17,987	18,902	19,901	20,466	21,637	21,634	23,007	23,186
Tenured and tenure-track faculty	78.2%	73.4%	68.9%	61.8%	60.8%	59.4%	59.7%	56.8%	55.7%	56.4%	55.5%	53.6%
Tenured faculty	53.1%	46.5%	48.1%	49.6%	48.5%	48.3%	47.6%	46.1%	45.3%	46.4%	45.1%	44.6%
Tenure-track faculty	25.0%	26.9%	20.8%	12.2%	12.3%	11.2%	12.2%	10.7%	10.4%	10.0%	10.4%	9.1%
Other academic positions	2.9%	4.7%	5.0%	9.4%	7.0%	8.4%	8.5%	9.5%	10.7%	9.9%	8.7%	7.7%
2-yr college faculty	0.7%	1.3%	1.8%	1.7%	1.5%	1.2%	1.2%	1.2%	1.2%	1.2%	2.0%	2.0%
Industry	8.7%	11.5%	13.3%	15.7%	18.2%	17.7%	19.1%	20.4%	20.1%	19.9%	20.2%	21.1%
Federal and other government positions	5.0%	4.3%	4.6%	5.5%	5.5%	4.7%	4.8%	4.3%	4.5%	5.0%	4.7%	4.5%
Self-employed and others	2.4%	3.2%	4.1%	4.1%	4.8%	6.3%	5.0%	4.6%	5.4%	6.0%	6.7%	6.9%
Postdoctoral appointments—other	0.3%	0.2%	0.2%	0.2%	0.5%	0.2%	0.2%	0.1%	0.2%	0.1%	0.2%	0.1%
Unemployed and seeking employment	1.4%	0.6%	1.1%	0.4%	0.6%	0.6%	0.4%	0.8%	0.4%	0.3%	0.9%	1.3%
Elementary- and high school teachers	0.2%	0.2%	0.3%	0.3%	0.6%	0.9%	0.7%	1.0%	0.8%	0.9%	0.5%	0.7%

Source: Analysis conducted by the National Research Council's Office of Scientific and Engineering Personnel for this study.

TABLE B-2 Occupation Status of PhD Mathematicians in the United States

	1973	1975	1977	1979	1981	1983	1985	1987	1989	1991	1993	1995
Research	2,605	3,100	3,637	4,272	4,483	4,620	5,411	5,409	5,958	6,005	5,848	5,902
Basic research	1,510	1,490	1,662	2,017	1,878	1,924	2,746	3,025	3,270	2,984	2,797	2,836
Applied research	799	919	1,222	1,321	1,588	1,335	1,444	1,799	1,921	2,299	2,676	2,660
Development	296	691	753	934	1,017	1,361	1,221	585	767	722	375	406
Research management	521	631	593	1,160	824	1,059	1,135	1,077	1,156	771	2,300	1,861
Management—other	665	932	1,265	1,357	1,230	1,164	1,660	1,360	1,398	1,438		
Teaching	7,454	8,622	8,603	8,601	9,311	9,590	9,460	9,484	9,794	9,161	10,239	10,108
Professional services	37	74	106	136	365	212	226	44	71	169	562	520
Consulting	177	222	232	406	648	763	611	448	629			
Computing										794	2,322	2,587
Other work activities, no response	299	414	907	748	927	1,389	1,312	2,547	2,542	2,571	1,525	1,907
Federal support	3,879	3,680	4,000	4,716	4,533	5,853	5,009	7,293	8,189	7,911	4,653	5,507
No federal support, no response	8,015	10,315	11,343	11,964	13,352	12,944	14,806	13,004	13,359	13,661	18,143	17,378
TOTAL	23,652	27,990	30,686	33,360	35,673	37,594	39,630	40,666	43,096	42,481	45,592	45,770
Research	11.0%	11.1%	11.9%	12.8%	12.6%	12.3%	13.7%	13.3%	13.8%	14.1%	12.8%	12.9%
Basic research	6.4%	5.3%	5.4%	6.0%	5.3%	5.1%	6.9%	7.4%	7.6%	7.0%	6.1%	6.2%
Applied research	3.4%	3.3%	4.0%	4.0%	4.5%	3.6%	3.6%	4.4%	4.5%	5.4%	5.9%	5.8%
Development	1.3%	2.5%	2.5%	2.8%	2.9%	3.6%	3.1%	1.4%	1.8%	1.7%	0.8%	0.9%
Research management	2.2%	2.3%	1.9%	3.5%	2.3%	2.8%	2.9%	2.6%	2.7%	1.8%	5.0%	4.1%
Management—other	2.8%	3.3%	4.1%	4.1%	3.4%	3.1%	4.2%	3.3%	3.2%	3.4%	0.0%	0.0%
Teaching	31.5%	30.8%	28.0%	25.8%	26.1%	25.5%	23.9%	23.3%	22.7%	21.6%	22.5%	22.1%
Professional services	0.2%	0.3%	0.3%	0.4%	1.0%	0.6%	0.6%	0.1%	0.2%	0.4%	1.2%	1.1%
Consulting	0.7%	0.8%	0.8%	1.2%	1.8%	2.0%	1.5%	1.1%	1.5%			
Computing										1.9%	5.1%	5.7%
Other work activities, no response	1.3%	1.5%	3.0%	2.2%	2.6%	3.7%	3.3%	6.3%	5.9%	6.1%	3.3%	4.2%
Federal support	16.4%	13.1%	13.0%	14.1%	12.7%	15.6%	12.6%	17.9%	19.0%	18.6%	10.2%	12.0%
No federal support, no response	33.9%	36.9%	37.0%	35.9%	37.4%	34.4%	37.4%	32.0%	31.0%	32.2%	39.8%	38.0%

Source: Analysis conducted by the National Research Council's Office of Scientific and Engineering Personnel for this study.

at different demographic and employment characteristics and by taking different cohorts. This provides for both longitudinal and time-series analyses, as shown here.

Of course, differentiating between research and teaching in determining the type of work for faculty is difficult. However it is fruitful to think about the nonresearch and teaching positions that mathematicians are obtaining and how they are changed over time.

Figure 4 in section 5.3.2 shows some of this information graphically. Note how the percentage of mathematicians employed as tenured and tenure-track faculty has declined while the percentage of mathematicians employed in industry has increased. The percentage in government employment has remained stable.

Figure B-8 shows the median salaries for PhD mathematicians and PhD holders in several related fields.

Figure B-9 shows the citizenship of faculty hired in 1991-1992 and figure B-10 the source of their PhDs.

Of particular concern is the unemployment status of new PhDs. Figure B-11 shows the change in unemployment rate for new mathematics PhDs from 1989 to 1996. The salaries of the new PhDs who attained academic employment are shown in figure B-12; the 9-month salaries included data on 102 men and 38 women, and the 12-month salaries included data on 20 men and 7 women.

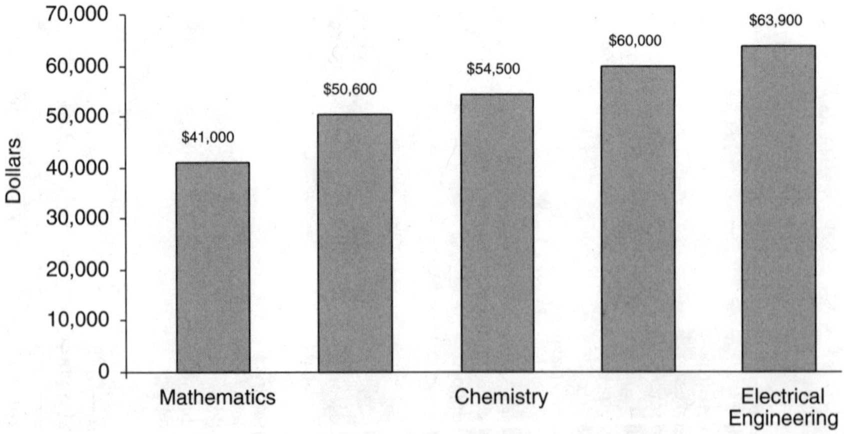

FIGURE B-8 Median salaries in 1993 of US PhDs who received their degree in 1985-1990, by field. Source: NSF 1996a, appendix table 5-27.

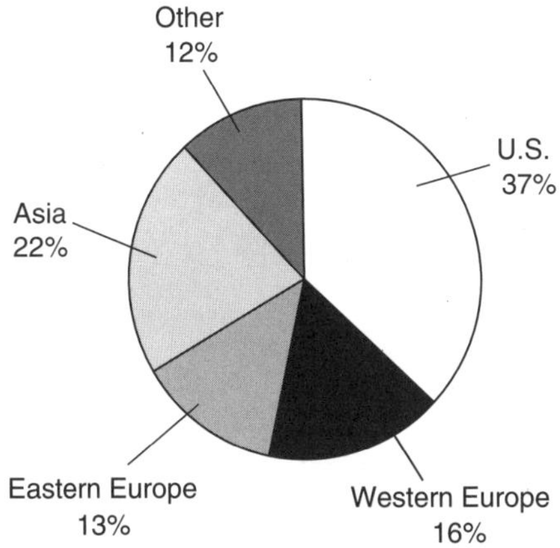

FIGURE B-9 Citizenship of full-time mathematics faculty with PhDs hired during 1991-1992 in the United States. Source: AMS 1992, pp. 314-315.

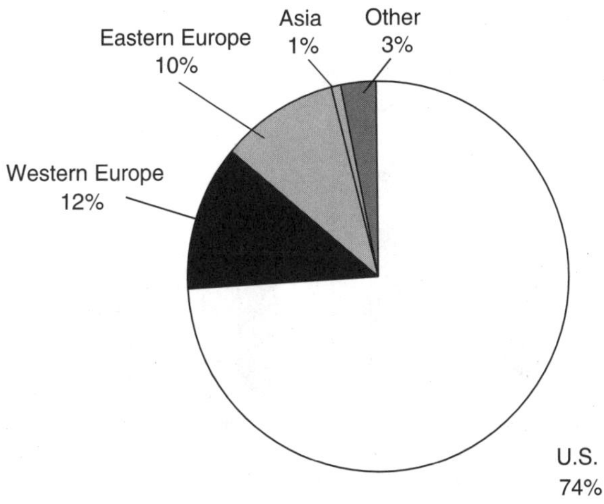

FIGURE B-10 Source of PhDs of full-time mathematics faculty hired during 1991-1992 in the United States. Source: AMS 1992, pp. 314-315.

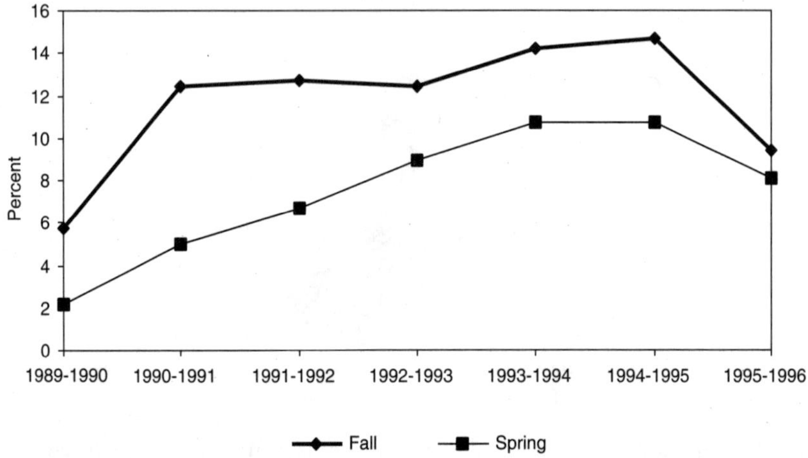

FIGURE B-11 Percentage of unemployed new US mathematics PhDs. Source: AMS 1996, 1997c.

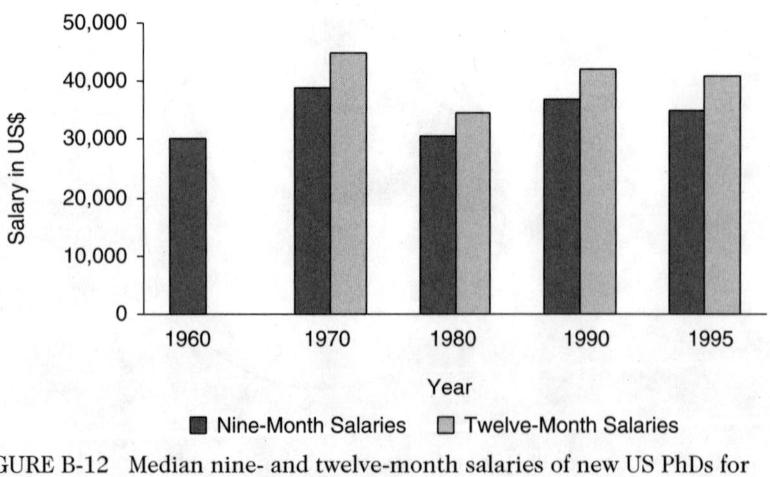

FIGURE B-12 Median nine- and twelve-month salaries of new US PhDs for teaching or teaching and research in 1995 dollars. Source: AMS 1996.

Funding

The information provided in this section, unless otherwise indicated, is from an analysis conducted by the Joint Policy Board for Mathematics for the American Association for the Advancement of Science. It produces an annual analysis of federal budget data on the field of mathematics.

There are 7 dedicated programs in mathematical sciences at 3 agencies: the Department of Defense (DOD), the Department of Energy (DOE), and the National Science Foundation (NSF). NSF focuses on fundamental research and its vitality, DOD looks on mathematical sciences as a problem-solving technology that can reduce costs in the development and deployment of hardware and software, and DOE and other agencies—such as the Department of Transportation, the Environmental Protection Agency, the National Aeronautics and Space Administration, the National Institutes of Health, and the National Institute of Standards and Technology—maintain mostly-applied mathematics and statistics activities to enable progress in fields related to their missions. All other agencies use applied mathematics and statistics.

Table B-3 shows federal support for academic mathematical-sciences research. Figure B-13 compares the percentage of academic mathematical scientists who have federal support to the percentages in other fields. Federal support for all mathematical research (basic, applied, and development) is shown in figure B-14.

The NSF Department of Mathematical Sciences (DMS) supports development of mathematical and statistical ideas and techniques, encourages the integration of mathematics with other disciplines, and encourages the diffusion of mathematics into technology. Grants are provided to individual investigators, research institutes, and centers for shared computing equipment, postdoctoral fellowships, research conferences, and undergraduate programs such as curriculum development.

NSF supports three mathematics institutes—the Institute for Mathematics and its Applications (IMA) at the University of Minneapolis was supported at $1,900,000 and the Mathematical Sciences Research Institute (MSRI) at the University of California, Berkeley was supported at $3,110,000 in FY1996. The IMA nearly matches the NSF support with funds from industry, sponsoring institutions, other agencies, and the University of Minnesota. The MSRI has limited additional support outside the NSF award. In 1998, there will be a recompetition for the location of the institutes in the mathematical sciences. The MSRI and the IMA are under review for "bridging" awards until the new national institutes are established as a result of the recompetition. Since its inception in 1989, the Center for Discrete Mathematics and Theoretical Computer Science (DIMACS) at Rutgers University and its staff have received a total of $74 million in science and technology center (STC)

TABLE B-3 Federal Support for the Mathematical Sciences, Fiscal Year 1995-1998, in Millions, Current Dollars

	Actual FY 95	Actual FY 96	Estimate FY 97	Percent Change[a] FY 96-97	Budget Request FY 98	Percent Change FY 97-98
National Science Foundation	87.69	91.70	98.22	7.11%	102.00	3.8%
DMS*	85.29	87.70	93.22	6.29%	97.00	4.1%
Other MPS	2.40	4.00	5.00	25.00%	5.00	0.0%
Department of Defense[b]	77.40	77.30	67.80	–14.01%	73.60	8.5%
AFOSR	17.50	16.70	17.10	2.39%	17.10	0.0%
ARO	15.00	15.00	13.00	–15.38%	15.00	15.4%
DARPA	21.00	22.90	19.50	–17.43%	22.40	14.8%
NSA	2.50	2.50	2.10	–19.05%	2.10	0.0%
ONR	21.40	20.20	16.10	–25.47%	17.00	5.6%
Department of Energy	15.70	16.00	16.00	0.00%	16.00	0.0%
University support	6.20	5.50	5.00	–10.00%	5.00	0.0%
National laboratories	9.50	10.50	11.00	4.76%	11.00	0.0%
TOTAL, All Agencies	180.79	185.00	182.02	–1.61%	191.60	5.3%

Federal Support for the Mathematical Sciences, Fiscal year 1995-1998, in Millions, Constant 1992 Dollars

	Actual FY 95	Actual FY 96	Estimate FY 97	Percent Change[a] FY 96-97	Budget Request FY 98	Percent Change FY 97-98
National Science Foundation	81.48	83.44	87.20	4.51%	88.26	1.2%
DMS*	79.25	79.80	82.76	3.71%	83.93	1.4%
Other MPS	2.23	3.64	4.44	1.99%	4.33	–2.5%
Department of Defense[b]	71.92	70.34	60.19	–16.86%	63.68	5.8%
AFOSR	16.26	15.20	15.18	–0.13%	14.80	–2.5%
ARO	13.94	13.65	11.54	–18.28%	12.98	12.5%
DARPA	19.51	20.84	17.31	–20.39%	19.38	12.0%
NSA	2.32	2.27	1.86	–22.04%	1.82	–2.5%
ONR	19.88	18.38	14.29	–28.62%	14.00	2.9%
Department of Energy	14.59	14.56	14.20	–2.54%	13.84	–2.5%
University support	5.76	5.00	4.44	–12.61%	4.33	–2.5%
National laboratories	8.83	9.55	9.77	2.30%	9.52	–2.5%
TOTAL, All Agencies	167.99	168.34	161.59	–4.18%	165.78	2.6%

[a] Column added by authors of this report.
[b] The FY1998 budgets for DOD's mathematical programs are estimates based on DOD's overall budget request for basic research.
*MPS = Directorate for Math and Physical Sciences.
Source: AAAS Report XII: Research and Development, FY 1998, Chapter 20

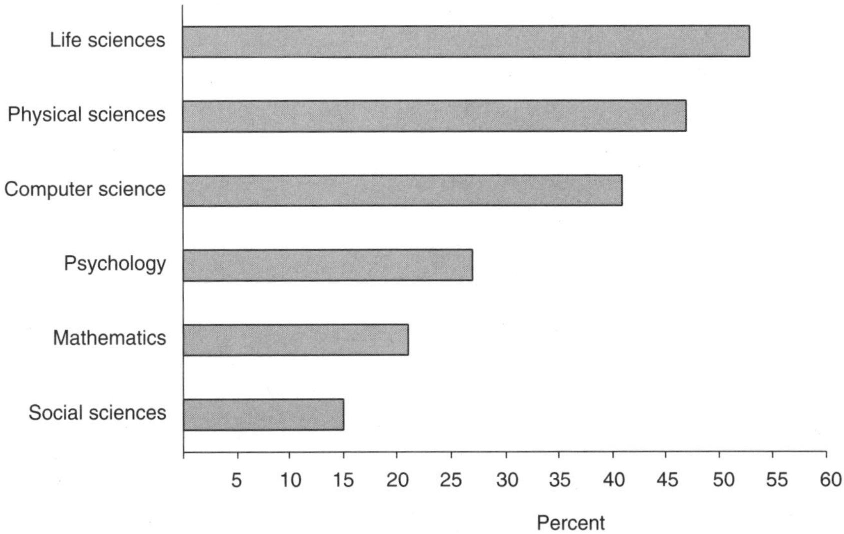

FIGURE B-13 Percentages of US academic scientists with federal support, 1993. Source: NSB 1996, appendix table 5-27.

and individual-investigator grants, of which NSF support has accounted for 50%. In 1995, total funding was $9.9 million. The STC program is nearing its end, and DIMACS will need to decide soon whether it will recompete for NSF STC funds. Other large projects supported by NSF include the Institute for Advanced Studies at $1,333,000 and the National Institute for Statistical Science at $1,068,000 in FY1996.

In DOD, the Air Force Office of Scientific Research supports research in subjects such as optimization, signal-processing, probability and statistics, computational mathematics, and dynamics and control. The Army Research Office focuses on the mathematics of materials science, high-performance computing, stochastic methods in image analysis, and mathematical and computational issues in intelligent manufacturing. The Office of Naval Research supports research in the mathematical subfields of applied analysis, discrete mathematics, numerical analysis, operations research, and probability and statistics. The Defense Advanced Research Projects Agency supports research that facilitates the development of technologies needed to meet future military needs. Of particular interest recently have been mathematical aspects of signal- and image-processing, electromagnetics, modeling and simulation of manufacturing processes, and optimized portable application libraries.

The National Security Agency is the nation's largest employer of mathematical scientists. It has a competitive grants program that

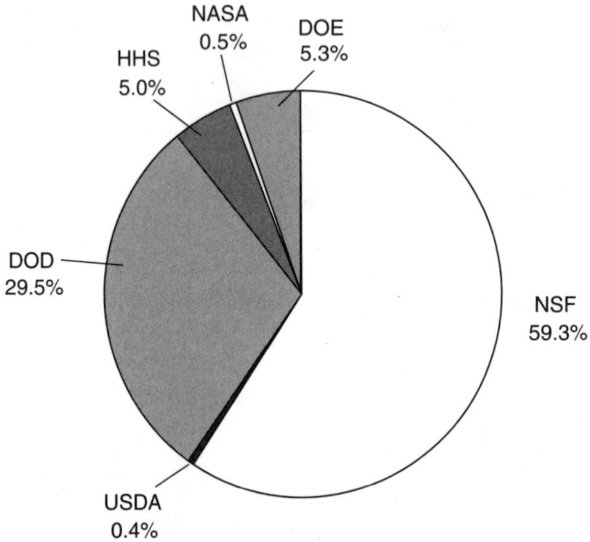

FIGURE B-14a Federal funding of US mathematical research—academic, 1993-1995 average.

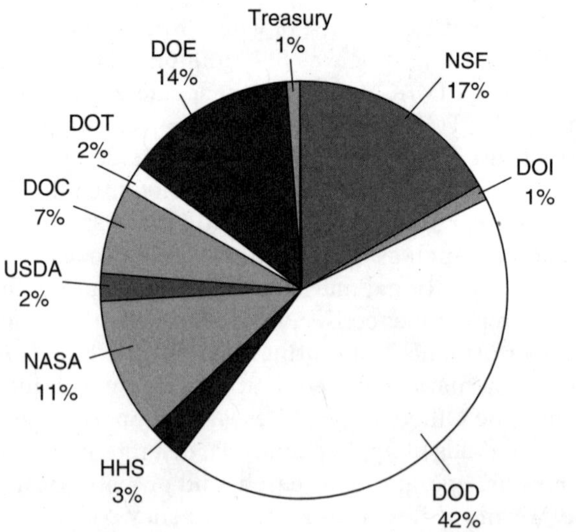

FIGURE B-14b Federal funding of US mathematical research—all R&D. Key: NSF = National Science Foundation; DOD = Department of Defense; DOE = Department of Energy; HHS = Department of Health and Human Services; NASA = National Aeronautics and Space Administration; USDA = Department of Agriculture; DOC = Department of Commerce; DOT = Department of Transportation; DOI = Department of the Interior. Source: NSB 1996.

supports unclassified academic research in discrete mathematics, algebra, number theory, probability, statistics, and cryptology.

The DOE focuses its R&D support on applied computer and computational mathematics, science and technology.

Papers and Citations

Two recent reports—one from Australia and the other from the United Kingdom—have analyzed scientific performance on a comparative basis using research-paper production and citation data. As noted in the Australian Bureau of Industry Economics report *Australian Science: Performance from Published Papers (1996)*, there are a number of problems in using such data, including a bias toward roman script and English-language journals; the greater attention paid to papers by renowned authors than to high-quality papers by less-known authors, technical papers, review articles, and recipes with little frontier science; and self-citation and citation circles.

Other problems occur because journal prestige and variation among disciplines is not considered. Time lag is a problem. There can be differential counting or miscounting due to multiple authorship, multiple field allocation, limits on the number of citations by journal, and changes in the number of journals in the field over time. And authors might use the same material with slight elaborations or break up a major article into several minor ones.

Papers "ahead of their time" and research communicated in nonjournal form (such as working papers, scientific equipment, computer programs, and seminar papers) might not be cited. Other outputs (such as teaching, advice to government, commercial research, and scientific services) are not included in bibliometric analyses.

Thus, citation rates measure visibility but not inaccessible work and not necessarily quality.

Figure B-15 shows the percentage of mathematical-research papers published by US authors relative to authors in 4 other countries that have strong mathematics programs. Figure B-16 compares the number of papers produced by US mathematicians with those produced in the European Community.

The UK report *The Quality of the UK Science Base* (1997) identifies the following as the top countries according to share of world's citations in mathematics:

1. United States.
2. United Kingdom.
3. Germany.
4. France.
5. Japan.

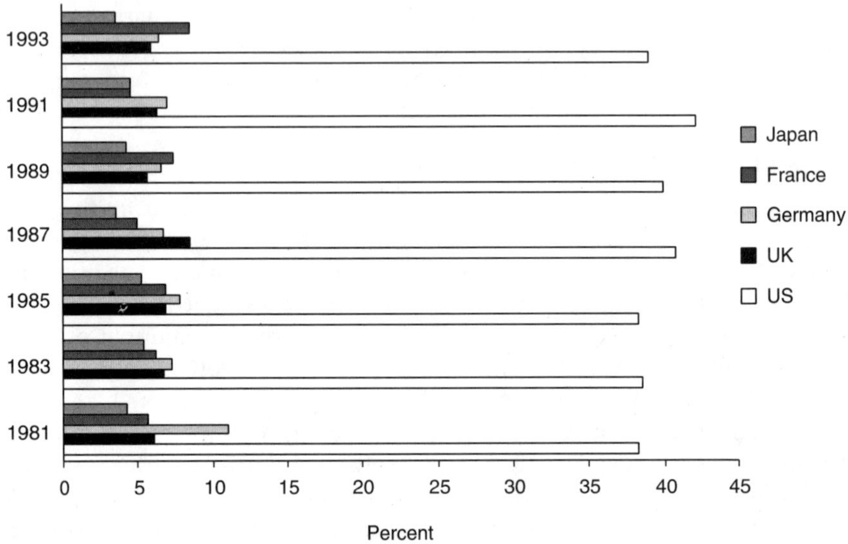

FIGURE B-15 Percentage of mathematical-research papers published by US authors. Source: NSB 1996, appendix table 5-31.

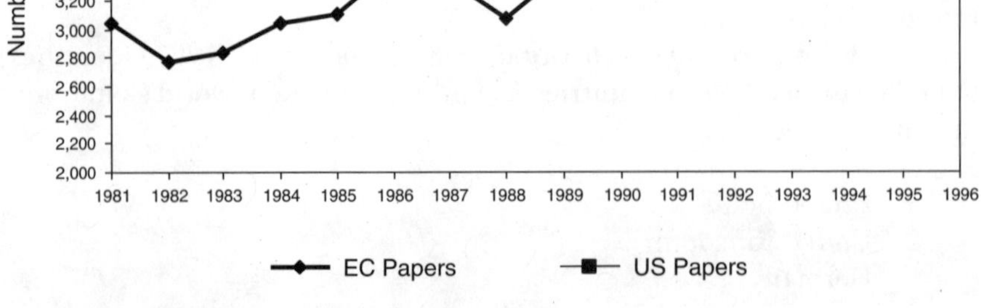

FIGURE B-16 Number of mathematical-research papers by US and EC authors, 1981-1996. Source: Institute for Scientific Information, *National Science Indicators on Diskette, 1981-1996*. Philadelphia, PA.

Another measure that was used in the UK report is the relative citation impact. The relative citation impact for a country in a particular field is defined as the country's share of the world's citations in the field divided by its share of world publications in the field. It can be thought of as a comparison of a country's citation rate for a particular field with the world's citation rate for the field. A relative citation impact (or rate) higher than 1 shows that the country's citation for the field is higher than the world's. According to the UK report, it is a measure of both the impact and the visibility of a country's research (as disseminated through publications) and gives some indication of the quality of the average paper.

The top countries in mathematics according to the relative citation impact index are:

6. Denmark.
7. Norway.
8. UK.
9. US.
10. Netherlands.

ATTACHMENT 2

INTERNATIONAL BENCHMARKING OF US MATERIALS SCIENCE AND ENGINEERING RESEARCH

Panel on International Benchmarking of
US Materials Science and Engineering Research

Committee on Science, Engineering, and Public Policy

International Benchmarking of US Materials Science and Engineering Research Panel Members

ARDEN L. BEMENT, Jr. *(Chair)*, Purdue University, School of Materials and Electrical Engineering, West Lafayette, IN
PETER R. BRIDENBAUGH, Executive Vice President, Automotive, Alcoa Technical Center, Alcoa Center, PA
LEROY L. CHANG, Hong Kong University of Science and Technology, Hong Kong
DANIEL S. CHEMLA, Lawrence Berkeley National Laboratory, University of California, Berkeley, Berkeley, CA
UMA CHOWDHRY, Business Planning and Technology Director, DuPont Specialty Chemicals, Wilmington, DE
ANTHONY G. EVANS, Gordon McKay Professor of Materials Engineering, Harvard University, Physics Department, Cambridge, MA
PAUL HAGENMULLER, Professor, Université de Bordeaux I, Laboratoire de Chimie du Solide du CNRS, France
JAMES W. MITCHELL, Director, Materials, Reliability and Ecology Research, Bell Laboratories, Lucent Technologies, Murray Hill, NJ
DONALD R. PAUL, Melvin H. Gertz Regents Chair in Chemical Engineering, Director, Center for Polymer Research, Department of Chemical Engineering, University of Texas at Austin, Austin, TX
BUDDY D. RATNER, University of Washington, Center for Engineered Biomaterials - UWEB, Seattle, WA
KATHLEEN C. TAYLOR, Head, Physics and Physical Chemistry Department, General Motors Corporation, GM Research and Development Center, Warren, MI
ROBERT M. WHITE, Professor & Head, Department of Electrical & Computer Eng., Carnegie Mellon University, Pittsburgh, PA
MASAHARU YAMAGUCHI, Professor, Kyoto University, Department of Materials Science & Engineering, Japan

Project Staff

DEBORAH D. STINE, Study Director
PATRICK P. SEVCIK, Research Associate
KATE KELLY, Editor

Materials Science And Engineering Benchmarking Guidance Group

MILDRED S. DRESSELHAUS *(Chair)*, Institute Professor of Electrical Engineering and Physics, Massachusetts Institute of Technology, Cambridge, MA

L. E. (SKIP) SCRIVEN, Regent's Professor, Department of Chemical Engineering and Materials Science, University of Minnesota, Minneapolis, MN

WILLIAM F. BRINKMAN, Vice President, Physical Sciences Research, Lucent Technologies, Murray Hill, NJ

GABOR A. SOMORJAI, Professor of Chemistry, Department of Chemistry, University of California at Berkeley, Berkeley, CA

ROBERT A. LAUDISE, Adjunct Chemical Director, Bell Laboratories, Lucent Technologies, Murray Hill, NJ

JAMES C. WILLIAMS, General Manager, Engineering Materials Technology Laboratories, GE Aircraft Engines, Cincinnati, OH

ALBERT NARATH, President, Energy and Environment Sector, Lockheed-Martin Corporation, Albuquerque, NM

PREFACE

In 1993, the Committee on Science, Engineering, and Public Policy (COSEPUP) of the National Academy of Sciences, the National Academy of Engineering, and the Institute of Medicine issued the report *Science, Technology, and the Federal Government: National Goals for a New Era*. In that report, COSEPUP suggested that the United States adopt the principle of being among the world leaders in all major fields of science so that it can quickly apply and extend advances in science wherever they occur. In addition, the report recommended that the United States maintain clear leadership in fields that are tied to national objectives, that capture the imagination of society, or that have multiplicative effect on other scientific advances. These recommendations were reiterated in another Academy report, *Allocating Federal Funds for Science and Technology*, by a committee chaired by Frank Press.

To measure international leadership, the reports recommended the establishment of independent panels that would conduct comparative international assessments of scientific accomplishments of particular research fields. COSEPUP indicated that these panels should consist of researchers who work in the specific fields under review (both from the United States and abroad), people who work in closely related fields, and research users who follow the fields closely.

To test the feasibility of that recommendation, COSEPUP is conducting experimental evaluations of three fields: mathematics, materials science, engineering, and immunology. The panel for each field has been asked to address the following three questions:

- What is the position of the United States in research in the field relative to that in other regions or countries?
- What key factors influence relative US performance in the field?
- On the basis of current trends in the United States and abroad, what will be the relative US position in the near term and the longer term?

Panels were asked to develop findings and conclusions, not recommendations.

This document provides the second of these assessments—that of the field of materials science and engineering. The panel found that it is critical that the United States lead the world in materials science and engineering innovations; however, the United States is not the leader in the field as a whole. Rather, it is among the world leaders in all subfields of materials science and engineering research and is the leader in some fields.

The panel found that the key to the nation's leadership is the flexibility of the materials science and engineering research enterprise, its innovation system, and its intellectual diversity. But, the ability of the United States to capitalize on its leadership opportunities could be curtailed because of shifting federal and industry priorities, a potential reduction in access to foreign talent, and deteriorating facilities of natural materials characterization. Of particular concern is the lack of adequate funding to modernize major research facilities in the United States when facilities here are much older than in other countries.

Once all the assessments are completed, COSEPUP will discuss the feasibility and utility of the benchmarking process and make whatever recommendations it deems necessary.

The committee thanks the panel for its hard work. We would also like to acknowledge those who made presentations at the panel meeting:

Steven Wax, Asst. Director for Materials and Processing, Defense Advanced Research Projects Agency

John J. Rush, NIST Center for Neutron Research, National Institutes of Science and Technology

Andrew J. Lovinger, Program Director, Polymers and NSF-wide Coordinator, Advanced Materials & Processing, National Science Foundation

This report has been reviewed by individuals chosen for their diverse perspectives and technical expertise, in accordance with procedures approved by the NRC's Report Review Committee. The purpose of this independent review is to provide candid and critical comments that will assist the authors and COSEPUP in making the published report as sound as possible and to ensure that the report meets institutional standards for objectivity, evidence, and responsiveness to the study charge. The content of the review comments and draft manuscript remain confidential to protect the integrity of the deliberative process. We wish to thank the following individuals for their participation in the review of this report:

John Armor, Principal Research Associate and Group Head/catalysis, Corporate Science Center, Air Products and Chemicals
Dan Drucker, Graduate Research Professor of Engineering Sciences, Emeritus, University of Florida
Merton Flemings, Toyota Professor, Massachusetts Institute of Technology
Lambert Ben Freund, Henry Ledyard Goddard University Professor, Division of Engineering, Brown University
Elsa Garmire, Dean, Thayer School of Engineering, Dartmouth College
William G. Howard, Independent Consultant, Scottsdale, AZ
Venkatesh Narayanamurti, Richard A. Auhll Professor and Dean of Engineering, University of Californa-Santa Barbara
William Nix, Lee Osterson Professor of Engineering and Professor of Materals Science and Engineering, Stanford University
William Spencer, CEO and Chairman, SEMATECH
Matthew Tirrell, Professor and Head, Department of Chemical Engineering and Materials Science and Director, Biomedical Engineering Institute, University of Minnesota
Jerry Woodall, Charles William Harrison Distinguished Professor of Microelectronics, Purdue University

While the individuals listed above have provided many constructive comments and suggestions, responsibility for the final content of this report rests solely with the authoring committee and COSEPUP.

Finally, the project was aided by the invaluable help of COSEPUP professional staff—Deborah D. Stine, study director, and Patrick P. Sevcik, research associate.

Phillip A. Griffiths
Chair
Committee on Science, Engineering, and Public Policy

CONTENTS

Figures, Tables, and Boxes

EXECUTIVE SUMMARY

To be leaders in industrial growth and to maintain a vibrant economy, it is critical that the United States lead the world in materials science and engineering innovations. Materials have been central to economic growth and societal advancement since the dawn of history. With the ever strengthening fundamental underpinnings of the fields and the growing interdependence of materials with other emerging technologies, these societal and economic contributions of the field are accelerating.

The Committee on Science, Engineering, and Public Policy (COSEPUP) Panel on International Benchmarking of US Materials Science and Engineering Research examined the leadership status of the United States in materials science and engineering research. Its members determined that the United States is among the world leaders in all subfields of materials science and engineering research and is the leader is some subfields, although not in the field as a whole. A general area of US weakness for most subfields is in materials synthesis and processing. Increasingly, US researchers must rely on specialty materials suppliers in Europe and Japan for bulk crystals and other specialty materials.

The United States is currently the clear leader in biomaterials and the leader in metals and electronic–photonic materials. However, the lead in electronic–photonic materials is endangered because industrial exploratory research has been cut back. The United States is currently one of several leaders in magnetic materials; previously, the United States had been preeminent, and this field needs particular attention in the future. US leadership is likely to be eroded in composites, catalysts, polymers, and biomaterials because of the high priorities given to these subfields by other countries. Of particular concern is the catalysts subfield, where there are not enough university multidisciplinary centers to conduct cutting-edge research and reduce the development cycle time for commercialization.

The panel also found that

- The flexibility of the materials science and engineering research enterprise is as much an indicator of its success as is its funding level.
- A major determinant of the nation's leadership in materials science and engineering leadership is its innovation system—the entrepreneurship ability of its researchers and the influence of its diverse economy.
- The nation enjoys strength in materials science and engineering through intellectual diversity—its ability to draw intellectually from all of the science and engineering research infrastructure.
- The ability of the United States to capitalize on its leadership opportunities could be curtailed because of shifting federal and industry priorities, a potential reduction in access to foreign talent, and deteriorating facilities for natural materials characterization. Of particular concern is a lack of adequate funding to modernize major research facilities in the United States—many are much older than are those in other countries—and to plan and build new facilities needed to maintain research leadership.

1

BACKGROUND

In 1993, the Committee on Science, Engineering, and Public Policy (COSEPUP) of the National Academy of Sciences, the National Academy of Engineering, and the Institute of Medicine issued the report *Science, Technology, and the Federal Government: National Goals for a New Era*. This report recommended that the United States be among the world leaders in all major fields of science to rapidly exploit exciting new concepts discovered elsewhere in the world. The report also says the country should maintain clear leadership in selected fields where achieving national objectives is critical or where public interest is acute. A similar recommendation was made in a later National Research Council report, *Allocating Federal Funds for Science and Technology*, published in 1995: The United States should "strive for clear leadership in the most promising areas of science and technology and those deemed most important to our national goals."

Both reports state that quantitative measures, such as dollars spent and number of scientists supported, are inadequate indicators of leadership and that policy decisions about programmatic issues or resource allocation would be better informed by comparative international assessments. Independent, field-specific panels were suggested as the best means to obtain such evaluations. Each panel would consist of researchers in the field, researchers in closely related fields, and research users who follow the field; each panel would include researchers from outside the United States.

In late 1996, COSEPUP began an experimental study of the effectiveness and outcome of such panels. The first panel report entitled

International Benchmarking of US Mathematics Research was released in October 1997. This report—an evaluation of US research in materials science and engineering—was prepared by the second panel. A study of the field of immunology is in progress. Each panel has been asked to address several questions:

- What is the position of US research in the field relative to that of other regions or countries?
- What key factors influence relative US performance in the field?
- On the basis of current trends in the United States and abroad, what will be the relative US position in the near term and in the longer term?

2

INTRODUCTION

2.1 How Important Is It for the United States to Lead in Materials Science and Engineering?

Materials are the substances from which things are or can be made. Materials science and engineering—the study of how to make, use, and adapt substances—has been central to social advancement and economic growth since the dawn of history. There has been an explosion in our understanding and application of materials science and engineering since the end of World War II, and the connection has become stronger between this field and other areas of emerging technology. The result has been an acceleration in the recent past of its contributions to social advancement and economic growth.

Federally-funded research on materials originally focused on defense and nuclear applications, but was expanded in the 1960s to include the space program, the protection of the environment, and the development of new energy systems. Today, research addresses issues in agriculture, health, information and communication, infrastructure and construction, and transportation. The future holds the promise of "intelligent" materials that will enable diverse technologies to respond dynamically to changes in the environment. A new class of materials, nanostructures, is already being used to advance the study of electromagnetics and mechanical properties (OSTP 1993).

Our national defense will continue to depend on providing the most advanced weapons to our military forces. Advanced materials are crucial to the improved performance and reliability of our weapons. Maintaining world leadership in materials essential to the design and manufacture of weapons will have high national priority.

To be leaders in industrial growth and to promote a vibrant economy, it is critical that the United States be among the world's leaders in all the subfields of materials science and engineering research. We need to be able to evaluate, adapt, and integrate materials identified and developed elsewhere in the world for use in new products and processes. Having world-class researchers who are knowledgeable about the frontiers of materials science and engineering is crucial to the rapid commercial assimilation and exploitation of important discoveries. Innovations in materials science abound in nearly all sectors of our economy. In agriculture, advanced natural polymers can be made from renewable resources that biodegrade more rapidly than plastics do. In energy, new materials and processing can be used to reduce energy costs significantly and conserve resources in the generation, transmission, and storage of energy. In protecting the environment, there is an opportunity to develop materials and processes that lead to cycles of infinite reuse. In health, biomaterials can be used to make artificial organs, joints, and heart valves; pacemakers; and lens implants, among others. Improvements in biomaterials could help improve the delivery of health care and reduce costs through the custom design of artificial biologic implants, for example, that will last a lifetime rather than a few years. In information and communications research involving semiconductors, new methods of design and processing could enhance the viability of the US electronics industry and open new marketing opportunities. In infrastructure and construction, the use of new or improved could reduce expensive maintenance of such structures as buildings, highways, bridges, and airport runways, to name a few. In transportation, materials research can help maintain US leadership in the increasingly competitive world aircraft market and reduce imports of oil and automobiles (OSTP 1993).

It is now possible to synthesize new materials atom by atom. The number of possible combinations of atomic assemblies to achieve new structures and properties is seemingly unbounded. But if the United States is to exploit these possibilities, strong national research capabilities by single investigators and multidisciplinary teams are required. Equipment—large-scale research instrumentation—will be required to characterize new materials from their smallest constituents at all scales of assembly. Computational methods are needed to find the best materials for a particular use.

Strengths in materials science and engineering research and education at US universities and colleges support other disciplines of science and engineering. The benefit increases with the growing unification of the field when multidisciplinary research can be done in centralized laboratories. Collaborative research benefits everyone because it helps identify new areas of endeavor and expand existing ones.

2.2 What Is Materials Science and Engineering?

The field of materials science and engineering research seeks to explain and control one or more of four basic elements:

- The properties or phenomena of a material that make it interesting or useful;
- The performance of a material; that is, the measurement of its usefulness in actual conditions of application;
- The structure and composition of a material, including the type of atoms that determine its properties and performance and their arrangement; and
- The synthesis and processing by which the particular arrangements of atoms are achieved (NRC 1989).

For the purposes of this report, the Panel divided materials science and engineering into 9 major subfields:

- Biomaterials
- Ceramics
- Composites
- Magnetic materials
- Metals
- Electronic and optical–photonic materials
- Superconducting materials
- Polymers
- Catalysts

These fields are described in Table 2.1 (modified from OSTP, 1993). The Panel has added the subfield of catalysts to OSTP's original list and combined two of the subfields—electronic and optical–photonic materials. It is important to appreciate that the classifications are arbitrary and overlapping. For example, supertough materials based on abalone shell biomimetrics are both biomaterials and composites. Figure 2.1 illustrates the interrelationships among categories.

2.3 What Key Factors Characterize the Field?

Materials science and engineering is multidisciplinary. Nearly all of science and engineering are involved in some way with some aspect of materials; the field involves internal and external interactions with the science and engineering communities at large. Scientists and engineers in many disciplines, including solid-state physics, chemistry, electronics, biology, and mechanics—not just those with materials science and engineering degrees—provide many of the ideas and motivation for materials science and engineering research.

TABLE 2.1 Materials Subfields

Biomaterials and biomolecular materials: Diverse materials compatible with human tissues or that mimic biologic phenomena; materials made from products of biologic origin. Traditional materials include dental fillings and crowns. Advanced materials are made from metals, ceramics, fibers, polymers, and natural biomolecules. Widespread applications are possible: artificial hearts, ultra–tough ceramic tank armor modeled on the molecular structure of abalone shells, biodegradable plastics for packaging, and nanofabricated circuit patterns on silicon for living neurons.

Metals: Tough, strong structural materials and electrical conductors. Traditional metals include commodity alloys of elements such as iron, nickel, and aluminum. Advanced metals tailored for specialty application include light-weight magnesium alloys; specialty tool steels and nickel-based alloys; refractory alloys; and high-temperature–high-strength intermetallics.

Ceramics: Materials made from nonmetallic inorganic minerals. Ceramics are noted for their light weight, hardness, and resistance to corrosion and high temperatures. Spark plug insulators are a traditional example. Advanced ceramics are used for thermal coatings and in high-temperature engines.

Superconducting materials: Materials that carry electrical current with no resistance. Some metals and alloys exhibit this characteristic but only at temperatures approaching absolute zero. Advanced varieties, including oxides, organics, and some intermetallics, superconduct at higher temperatures, some exceeding the liquefaction temperature of nitrogen.

Composites: Hybrids of at least 2 materials, usually reinforced ceramics, metals, or organic matrix materials, which are combined to exploit the most useful properties of each. Fiberglass is a traditional composite, composed of glass fibers in an epoxy matrix. Advanced composites have structural and nonstructural applications and often are used in air and land vehicles.

Polymers: Large molecules consisting of long chains of repeated units. Polymers are noted for unique combinations of properties and have a range of applications, from plastic containers to liquid crystal displays. Plastic wrap is a traditional example. Polyimides are advanced, high-temperature polymers used for electronic packaging and aircraft skins.

Electronic materials: Electronic materials are active materials, such as semiconductors, that transmit signals by way of electrons. Current electronic technology is based on silicon but, newer semiconductors include compound semiconductors, (gallium arsenide), wide-band-gap semiconductors (silicon carbide). These compound semiconductors also are considered optical–photonic materials (see box). This class includes metals, ceramics, and polymers used in electronic wiring, interconnections, and packaging.

Optical–photonic materials: Materials that transmit light; those used as light sources, such as lasers; and those used to switch and modulate light. Glass is in this category in numerous forms, from window panes to optical fibers.

continues

TABLE 2.1 Continued

Magnetic materials: Materials that possess spontaneous magnetization. Their magnetic fields make them useful in motors and generators; the orientation of magnetization can be used to store information. Magnetic materials can be metallic, such as iron and iron–rare earth alloys, or nonmetallic, such as oxides.	***Catalysts:*** Materials that accelerate chemical reactions without being consumed in the process. Catalysts find wide use for production of chemicals and pharmaceuticals, refining of petroleum, and for control of emissions of the products of combustion (for example, from motor vehicle engines). The benefits of catalytic processes include low process costs, improved productivity, high selectivity to desired product, and reduction of unwanted by-products.

Nearly all modern industries benefit from developments in materials research. Because there is considerable overlap in the study of materials problems among industries, solutions have enormous economic leverage. Semiconductors, for example, are at the foundation of the electronics industry. The development of new materials also has a large economic multiplying effect because it creates demands for new processing equipment and manufacturing tools.

Research in materials science and engineering is capital intensive and involves increasingly sophisticated characterization instruments and equipment for synthesis, processing, and analysis. The equipment ranges from small, laboratory bench-scale machines that serve a single investigator to synchrotron sources, nuclear reactors, superconducting magnets, and supercomputers that serve large user communities and research groups. The field benefits from the large US installed base of research facilities.

Problems in materials science and engineering research require all forms of research, from small-scale research carried out by a principal investigator and a small team, to large multidisciplinary teams, and regional consortia involving many investigators. Consortia, alliances, and partnerships of industrial, university, and government laboratories are a common mode of exploiting breakthroughs in the field. Equally common are the international collaborations made possible by the explosive growth of the Internet.

Computational research and engineering, involving large-scale supercomputers and computer networks, is gaining importance in solving all manner of materials problems—from the subatomic to the macroscopic scale. Considerable progress has been made recently in simulations of complex materials phenomena based on first principles, such as mesophysical and mesomechanical phenomena. Computational

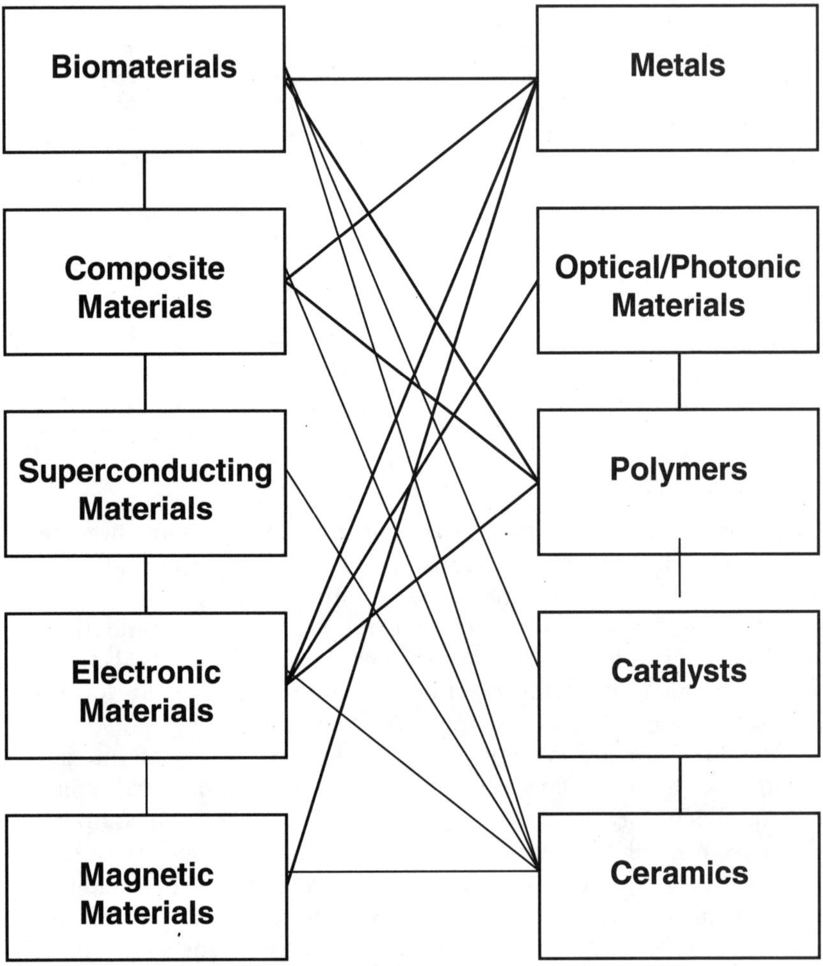

FIGURE 2.1 Inter-relationships among materials categories.

strategies are emerging to provide physical descriptions of materials over a range of sizes important to a given process. In some instances, the use of these strategies allows the prediction of system performance that is not possible now by direct measurement. The field benefits directly from US strengths in computer science and engineering.

New developments in materials science and engineering can aid rapid paradigm shifts in the development of new technologies in fields that are not directly related to materials research. For example, the discovery of high-temperature superconductors just a decade ago is leading to important technological developments in medicine, defense, energy, computing, telecommunications, and transportation. These developments will enable important expansions in the global market of the next century.

Because of the increasing severity of the environments in which many advanced materials are used, the time from first synthesis to practical, reliable application can be long, often fifteen years or more. Long-term research is expensive, so sustained public-sector investment in precompetitive research and development is critical for realizing the economic potential of new materials discoveries. Strong user involvement in the early stages of materials synthesis and applications research is critical for facilitating the early adoption of new materials for new or existing applications.

2.4 What Is the International Nature of Materials Science and Engineering?

Materials science and engineering is an international effort that affects an individual nation's economic, industrial, and military strength and the education of its citizens. Because of the importance of materials to economic strength and industrial success, most major US trading partners have targeted materials science and engineering as a growth area and have made major investments to build competency in the field. Materials science and engineering is prominently represented in national public–private sector partnerships for economic development in most European countries and in the Pacific Rim countries, notably Japan. National and multinational companies with strong research and development programs in materials technologies market their products worldwide.

Materials technology is critical to the development of advanced military weapons and is one determinant of military strength. Nations that supply military equipment, such as the US, the United Kingdom, Germany, France, and Russia, have built strong industrial and government laboratories that specialize in military-related materials research, although much of their results find civilian applications as well.

Leadership in the subfields of materials science and engineering can shift unexpectedly. Many prominent researchers in materials science and engineering around the world have received graduate education in US research universities. This facilitates international collaboration and exchange with US investigators. Most new discoveries are immediately communicated around the globe today, and most new materials developments are exploited in many countries simultaneously. Likewise, many new discoveries are now announced simultaneously by researchers in different countries.

2.5 What Are Some Caveats?

Because of the size and industrial strength of the US materials science and engineering research community, it cannot be compared meaningfully with those of other single countries. The only sensible method is to compare the US with regional groups, such as Europe or Asia, for example. To the extent possible, in this report, specific countries are mentioned in connection with particular areas of science and technology.

Because of the enormous breadth of the field, it is necessary to divide materials science and engineering research into subfields, each of which is also extremely broad. The panel adopted the material subfields identified by the White House Office of Science and Technology Policy (OSTP) National Science and Technology Council in its 1993 and 1995 reports, *The Federal Research and Development Program in Materials Science and Technology*. The reports list biomaterials, ceramics, composites, electronic materials, magnetic materials, metals, optical–photonic materials, polymers, and superconducting materials. The panel added catalysts to this list and combined the electronic and optical–photonic materials research into one category.

Fundamental materials discoveries can occur in many research settings. For example, NITINOL (a memory alloy) was discovered at a government laboratory, high-temperature superconductivity was discovered at an industrial laboratory, and rapidly solidified amorphous metals were first produced at a university. Many such developments occur in all three settings at the same time, leading to synergistic breakthroughs. Centers of excellence abound in all three settings in the United States as well as abroad.

Multidisciplinary research is a common mode for individual investigators as well as for large research teams. In industrial and governmental laboratories, materials research is, for the most part conducted by multidisciplinary teams, often led by scientists or engineers from diverse disciplinary backgrounds. For example, advanced polymer research is commonly led by chemists, and research on electronic materials is commonly led by physicists. Mathematicians often are involved in theoretical studies and in the development of computational models and simulations. At universities, most large science and engineering departments have self-contained research groups that focus on materials-related science, engineering, or both.

In the United States, there has been a strong unification of the field over the past 3 decades, to include the development at most universities of a unified curriculum across the field. At many research universities, the first course in materials is offered to freshmen. There are still many important materials-related courses provided outside materials science and engineering departments, so instruction also has interdisciplinary aspects. Although departments of materials science and engineering are often found in schools of engineering in the United States, they are commonly found in schools of science or natural history abroad.

2.6 Panel Charge and Rationale

The Panel was asked to conduct a comparative international assessment to answer three questions:

- What is the position of US research relative to that of other regions or countries?
- What key factors influence US performance in the field?
- On the basis of current trends in the United States and abroad, what will be the relative US position in the near term and in the longer term?

The panel was asked only to develop findings and conclusions—not recommendations. The panel focus is on leading-edge exploratory research—intermixing basic and applied research and product development.

The panel responded to the second question first, identifying the determinants of leadership that have influenced US advancement in the field and the establishment of the supporting research infrastructure. Section 3 of the report details the panel's findings.

The panel then assessed current US leadership in the nine subfields. The results of this assessment—the benchmarking results given in Section 4 of the report—are in response to the first charge.

The next step was to assimilate past leadership determinants and current benchmarking results to predict US leadership, thereby to address the third charge. This analysis is given in Section 5 of the report.

The panel next attempted to predict—based on near-term and longer term trends in the determinants of leadership and in corresponding developments around the world—leadership positions of the United States in the subfields of materials science and engineering. That is, would the United States gain, maintain, or lose position with respect to its current state? Section 6 of the report discusses the Panel's predictions for each of the subfields assessed. Tables in Appendix B provide specific analyses by sub-subfield.

The panel's principal findings and conclusions are given in Section 7.

3

DETERMINANTS OF SCIENTIFIC LEADERSHIP

Leadership in materials science and engineering is influenced by several factors, that would be weighted differently in highly developed countries around the world depending on national policy, economics, and available resources (installed research infrastructure, talent base). The Panel focused on 5 determinants as particularly important to US leadership in the various subfields of materials science and engineering:

- ***National imperatives:*** To what extent do national imperatives for defense, infrastructure development, or international competitiveness influence science and technology policy?
- ***Innovation:*** What investment and technology development mechanisms facilitate introduction of new materials and processes into the marketplace?
- ***Major facilities:*** What facilities exist to elucidate atomic and subatomic materials structure and phenomena?
- ***Centers:*** What research centers exist to facilitate interdisciplinary research?
- ***Human resources:*** What infrastructure exists to educate and train scientists and engineers in the various disciplines that support materials science and engineering? Does this infrastructure provide a tiered community of leaders for technology research and development?
- ***Funding:*** Are sources of support balanced and adequate to sustain leadership in the areas of research that support the national imperatives?

Although possession of these elements does not guarantee leadership, without a majority of them, leadership would be difficult to maintain. Based on panel assessments, the United States does currently enjoy a leadership position—or is among the leaders—in a majority of the subfields of materials science and technology. The panel focused on the status and trends of the determinants listed above and found that they have not only made the US position of leadership possible, but they also will be critical for sustaining leadership in the future. Although presenting a balanced, comprehensive, comparative analysis of these determinants for all of the regions of the world would take more time than is available for this study, comparisons are made where they are particularly salient and information is readily available.

3.1 National Imperatives

Several forces coalesced after World War II to bring about what, during the 1950s, was known in the United States as the National Materials Program. The global spread of nuclear weapons capabilities, the growing intensity of intercontinental ballistic missile development, and the space race placed demands on the materials science and engineering communities for advanced materials that would give the United States a strategic edge. Also in the 1950s, leading industrial laboratories had a growing interest in materials research and development and set up well-equipped interdisciplinary laboratories that contributed to economic growth.

The euphoria over developments in synthetic polymers, uses of the transistor, and optical fibers and the emergence of rudimentary composite materials spurred interest in the White House Science Office, the Office of Naval Research (ONR), the National Aeronautics and Space Administration (NASA), the Atomic Energy Commission (AEC), the National Science Foundation (NSF), and the Department of Defense (DOD) for the establishment in 1960 of what became known as the Defense Advanced Research Projects Agency (DARPA) Interdisciplinary Laboratories (IDLs). Later in 1972, the DARPA IDLs were transferred to NSF to become the Materials Research Laboratories (now the Materials Research Science and Engineering Centers). Also, beginning in the 1960s, AEC and NSF invested in major instrumentation facilities at national laboratories and research universities to probe more deeply the atomic and subatomic properties of matter.

The national imperative for materials research and development was sustained until the mid-1980s by the intensifying cold war, the space race, and growing concern over global competitiveness of US manufacturing. Materials have been named on all the various "critical technologies" lists, in particular those of the Department of Commerce, DOD, and Rand's Critical Technology Institute. Development of the infrastructure for materials research and development, the increased unification of the field by bringing together materials subdisciplines and by integrating science with engineering and theory with experimentation, is attributable to the momentum created by national imperatives. Thus national imperatives can be credited for much of our current leadership. The major contracts government provides to industry are also, to a large degree, in response to national imperatives. Federally funded programs have contributed to industrial leadership in the semiconductor, civilian aircraft, computer, and optical–electronic industries, among others. Therefore, an understandable concern in the materials research and development community is whether leadership can be sustained in the absence of nationally focused imperatives.

3.2 Innovation

Beauty and elegance can describe a scientific discovery, but its power lies in its effect on our lives. A key factor in its effect is how rapidly and easily ideas can be tested, developed, and extended. The process by which research ideas are developed and funded in the United States—our "innovation system"—is unique. The factors that influence the process—pluralism, partnerships, regulation, and professional societies—are discussed below.

3.2.1 Pluralism

The funding of our innovation system is characterized by many options, whether in academic research or in the entrepreneurial work supported by small and large companies. NSF and DOE also support the work. Programs are funded by mission agencies, such as DARPA, or the defense science offices (ONR, Army Research Office, and Air Force Office of Scientific Research). This variety of sources, with different emphases, creates a spectrum of opportunities. For example, DOD might be interested in giant magnetoresistance for radiation-hard nonvolatile memories; NSF might want to encourage collaboration in this area between universities and the magnetic storage industry. The peer review process that underlies research funding and the extensive networking associated with advisory boards contributes to the high quality of federally funded research.

In addition, large and small industries conduct basic and applied research independently. Industry needs influence funding by the federal government, and history shows that pluralism is an important influence of research leadership. Thus, the direction of research should never be dictated solely by industry.

3.2.2. Partnerships

Collaboration of university and industry researchers is an important aspect of the US innovation system. In other countries, such as

Germany, similar connections exist, and European economic development programs support these relationships. Many Japanese companies support materials research in US universities because of our ready access to new technology. Whereas US industry funds only about 10% of the research carried out in universities, the mobility of individuals between academic and industrial laboratories is especially vital in the transfer of new concepts and technology. Many faculty members have industrial experience, and often they serve as consultants to industry. University faculty also participate in the formation of high-tech companies. These relationships provide university researchers with an understanding of problems that are relevant to industry, and they provide a channel for the transfer of knowledge and new approaches developed in academia with funding from the federal government. There are few true two-way university–industry collaborations, for example, where industry funds the research and influences its direction through joint activities, such as two-way personnel exchanges.

A good example of one industry–university–government collaboration can be found in NSF's Engineering Research Centers. The data storage center at Carnegie-Mellon University in Pittsburgh is actively supported by the US industry, as represented by IBM, Seagate, Quantum, and others. One limiting factor in high-density recording has been the noise associated with thin-film magnetic media. Collaborative studies by industry and Carnegie-Mellon scientists have identified the source of the noise, and these studies have led to the development of a Nickel–Aluminum film underlayer that promotes the growth of low-noise media. The material is being adopted by the industry, and it will provide US companies with a competitive advantage.

Partnerships in general, whether between universities and industry or among companies, have become critical to improving the effectiveness with which industry commercializes research. Some corporations now rely as much on academic research as they do on work conducted in their own laboratories. Many researchers worry that the abandonment of forward-looking exploratory research by some large companies will lead to weakness in materials science and engineering research in the United States. This seems particularly true in cases where materials fabrication and characterization are capital intensive. Direct involvement of materials producer and user communities in the early stages of research and development at universities and national laboratories can be essential to the use of new materials technology in the design and manufacture of new products.

3.2.3 Regulation

Government regulations are another factor in the innovation process. The objectives of safety, and of environmental, occupational, and health protection can add time and expense to the development of new materials. In general, long lead times—sometimes exceeding 15 years—are required to bring a new material to the marketplace. This delay is caused not only for these objectives but also because of the time needed to develop processing methods, reduce property variations, develop design data bases, and develop standards for the testing and use of new materials. Government regulations also can motivate the need for materials. Examples include catalysts used to control hydrocarbon and nitrogen oxide emissions and on-board sensors required by the Clean Air Act for monitoring automobile emissions.

The materials subfields that are especially affected by regulation are biomaterials, metals, polymers, ceramics, and composites used in safety-critical structures (airframes, nuclear reactors, jet engines). Regulatory barriers that unduly extend lead times to market in these materials subfields can directly affect the US global position.

Environmental control regulations affect research on such materials as polymers, adhesives, and coatings, and in the study of volatile organic compounds, hazardous elements, and biodegradation.

3.2.4 Professional Societies

Another factor in achieving leadership is the information infrastructure provided by professional societies. The US materials science and engineering community has benefited from the diversity of professional societies, which facilitate communication, organize and focus attention on new topics via symposia, produce world-class journals and publications, and sponsor international conferences and workshops that bring researchers together (Box 3.1). Prominent professional societies in many disciplines also have materials-related divisions (the Institute of Electrical and Electronics Engineers [IEEE], the American Physical Society [APS], the American Chemical Society [ACS], the American Institute of Chemical Engineers [AIChE], the American Agricultural Economics Association [AAEA], and the American Society of Mechanical Engineers [ASME]). Many researchers in a variety of disciplines actively pursue materials research. The establishment of scholarly and professional societies in the United States is dynamic. The Materials Research Society (MRS) was established in 1973 to provide professional representation for materials scientists working on electronic, photonic, and other functional materials. MRS now has more than 12,000 members in the United States and in more than 50 other countries; it has been broadly emulated by other nations under a confederation of International Council of Scientific Unions (ICSU) materials societies (Box 3.2). Materials scientists and engineers are also prominently represented in learned societies here and abroad. For example, 2.8% of the members of the National Academy of Sciences (NAS) and 15.5% of the members of the National Academy of Engineering (NAE) can be identified with materials research. They also often enjoy cross-member-

BOX 3.1 The Federation of Materials Societies

The Federation of Materials Societies (FMS) is an umbrella organization whose member societies and affiliates represent professional societies, universities, and National Research Council organizations involved with materials science, engineering, and technology. FMS constituent societies have more than 700,000 members.

Constituent Societies
- American Association for Crystal Growth
- The American Ceramic Society, Inc.
- American Chemical Society
- American Institute of Chemical Engineers
- American Physical Society
- American Vacuum Society
- ASM International
- American Society of Mechanical Engineers International
- American Society for Nondestructive Testing
- American Society for Testing and Materials
- American Welding Society
- The Electrochemical Society, Inc.
- International Society for Hybrid Microelectronics
- Materials Research Society
- National Association of Corrosion Engineers International
- North American Catalysis Society
- The Minerals, Metals & Materials Society

Affiliates
- University Materials Council
- Conoco, Inc.
- Dow Corning, Inc.

Represented by Liaison Members
- National Materials Advisory Board
- Solid State Sciences Committee

Source: http://www.foms.org/

ship in learned societies around the world. Such associations and collaborations greatly facilitate awareness and global exchange of fast-breaking developments in the field.

3.3 Major Facilities

Research on materials depends on the ability to fabricate and characterize materials. To be a leader, one must have access to state-of-the-art facilities. An important component of federal support for basic

BOX 3.2 International Union of Materials Research Societies

The International Union of Materials Research Societies was established in 1991 as an association of technical groups and societies that have an interest in promoting interdisciplinary materials research.

The Union's objectives are as follows:

- To facilitate international cooperation among materials research organizations,
- To contribute to the advancement of materials research in all its aspects,
- To advance the multidisciplinary nature of materials research internationally,
- To promote information exchange among national or regional societies with interests in interdisciplinary materials research, and to work to coordinate their activities, and
- To promote communication of international materials research activities through appropriate media and to encourage well-established materials research symposia to rotate through available meeting sites of materials research societies.

Current members:

- Australian Materials Research Society
- Chinese Materials Research Society
- European Materials Research Society
- Materials Research Society
- Materials Research Society of India
- Materials Research Society of Japan
- Materials Research Society of Korea
- Materials Research Society of Russia
- Materials Research Society of Taiwan
- Mexican Materials Research Society

Source: http://mrcemis.ms.nwu.edu/iumrs

research is the development and maintenance of the major facilities needed to carry out that research. By "major facilities," we mean facilities that have unique research capabilities that are too expensive for any one entity to support. In the case of materials research, these facilities include sources of neutrons, synchrotron radiation, high-energy electrons, and high magnetic fields.

Major facilities serve as an intellectual focus, and science develops as the interplay between experiment and theory. State-of-the-art facilities attract the world's leading scientists, so excellent facilities often make for award-winning research. Table 3.1 lists awards given to scientists for their work in neutron-scattering research.

3.3.1 Neutron Scattering Facilities

Tables 3.2–3.5 show neutron sources in the US and the rest of the world. As shown in these tables, US sources, if not upgraded, soon will have less capability than found in sources abroad. Many US facilities also are oversubscribed. The National Institute of Standards and Technology (NIST) cold neutron facility receives up to 3 times as many research proposals as it can accommodate, and responses to proposals can take 6 months. The situation is similar at Argonne National Laboratory, where oversubscription is 2–2.5, occasionally as high as 7. The review time for proposals is 6 months at minimum, and some proposals are never considered because peer reviewers rate their quality as too low for inclusion (Personal communication, from Bruce Brown, ANL). It is not clear that Argonne National Laboratory's peer review standards are equivalent to those elsewhere.

3.3.2 Synchrotron Sources

Table 3.6 provides statistics on the synchrotron sources in the United States and other G-7 countries based on a recent report from a DOE advisory committee (DOE, 1997).

Synchrotron radiation (Ultraviolet and X-ray) is used in a variety of techniques (absorption, scattering, spectroscopy, and microscopy) to examine the intricate electronic, atomic, and geometric structures of many materials. Such diverse problems as magnetic phenomena in thin films, the chain structure in polymer blends, and the electronic and bond structure of catalysts are being examined by synchrotron sources. There is some concern within the synchrotron radiation source community that the development of third-generation sources in Europe (such as ESFR and Elettra) will attract users away from second-generation sources in the United States, although such a reduction in demand has not yet occurred. Third-generation sources will not necessarily improve many experiment-limiting factors, such as flux limitations, detector capability, and source stability. Therefore, some argue that useful science can be expected from US synchrotron facilities for some years to come (Hart 1997). Furthermore, in many cases, Americans have access to facilities in Europe.

The DOE report assesses the cost-effectiveness of this research at DOE facilities and makes recommendations on funding priorities for these facilities.

TABLE 3.1 Major Scientific Awards for, or Strongly Influenced by, Neutron-Scattering Research

Year	Name	Award	Research Area
1957	Clifford Shull (MIT)	APS Buckley Prize	Neutron diffraction, magnetic structure
1963	Bertram Brockhouse (AECL)	APS Buckley Prize	Phonons, magnons
1973	John Axe, Gen Shirane (BNL)	ACA Warren Diffraction Award	Soft modes, phase transitions
1973	Gen Shirane (BNL)	APS Buckley Prize	Phonons, soft modes
1974	Paul Flory (Cal Tech)	Nobel Prize, Chemistry	Polymer structure
1978	Henri Benoit (Strasbourg)	APS High Polymer Prize	Neutrons, polymer structure
1982	Edwards (Cambridge) and Pierre de Gennes (Col. Paris)	APS High Polymer Prize	Reptation theory
1984	Charles Han (NIST)	APS Dillon Medal	Polymer structure and dynamics
1986	Muthu Kumar (U. Mass.)	APS Dillon Medal	Theory of polymer structure
1987	Robert Birgeneau (MIT)	APS Buckley Prize	Magnetism
1988	Robert Birgeneau (MIT), Paul Horn (IBM)	ACA Warren Diffraction Award	Low-dimensional systems
1988	Jean Guenet (Saclay)	APS Dillon Medal	Gel formation
1989	Frank Bates (AT&T)	APS Dillon Medal	Block copolymers
1990	Pierre de Gennes (Col. Paris)	Nobel Prize	Theory of polymers, liquid crystals
1990	James Jorgensen (ANL)	ACA Warren Diffraction Award	Structure of ceramic superconductors
1990	Dieter Richter (KFA) and John Huang (Exxon)	Max Planck Research Prize	Dynamics of polymers and microemulsions
1991	Ken Schweitzer (Sandia)	APS Dillon Medal	Polymer RISM theory
1992	Glenn Frederickson (UCSB)	APS Dillon Medal	Theory of microsphere polymer structure
1992	Phil Pincus (UCSB)	APS High Polymer Prize	Theory of complex fluids
1992	Alice Gast (Stanford)	Colburn Award (American Institute of Chemical Engineering)	Colloids and polymers
1994	Schull and Brockhouse	Nobel Prize, Physics	Neutron-scattering methods for structures
1996	Frank Bates (U. Minn.)	APS High Polymer Prize	Structure of copolymers
1996	Nitash Balsara (N.Y. Polytech.)	APS Dillon Prize	Properties of polymer blocks
1997	David Price (ANL)	ACA Warren Prize	Structure of disordered systems

Source: Adapted from, *Neutron Sources for America's Future: Report of the Basic Energy Sciences Advisory Committee Panel on Neutron Sources*, Department of Energy, DOE/ER-0576P, 1993.

TABLE 3.2 US Research Reactors

Facility	Agency	Year	Thermal Flux/Power	Operation Cost ($ million FY 1996)
HFBR (BNL)[a]	DOE	1965	10^{15}/60 MW	25 (60 MW)
HFIR (ORNL)	DOE	1966	10^{15}/85 MW	27
HFIR upgrade		2001	Cold source 4 cold instruments	
NBSR (NIST)	DOC	1969	4.10^{14}/20 MW	7
NBSR upgrade		2000	New cold source (X2) 5 thermal/cold instruments	
MURR (U. Mo.)		1965	10^{14}/10 MW	6

[a] This facility is closed. Plans are to bring it up to 50% power 30 MW.
Source: Presentation by John Rush of NIST to Panel in 8/97.

3.3.3 *Nanofabrication*

Of particular interest to those involved in electronic materials is fabrication of nanostructures. For many years, university microelectronic designs have been tested by having MOSIS (the Metal-Oxide Semiconductor Implementation Service) build prototypes (Box 3.3). Although research and development programs for microelectromechan-

TABLE 3.3 Research Reactors Abroad

Facility	Year	Thermal Flux/Power	Operation Cost ($ million FY 1996)
ILL (France)	Refurb. 1995	$1.2.10^{15}$/57 MW	26
Orphée (France)	1980	3.10^{14}/15 MW	~16
KFA (Germany)	Refurb. 1994	2.10^{14}/23 MW	~17
Berlin (Germany)	Refurb. 1991	2.10^{14}/10 MW	~15
Riso (Denmark)	1963	$1.5.10^{14}$/12 MW	~12
JRR-3M (Japan)	1991	3.10^{14}/20 MW	18
HANARO (Korea)	1994	20 MW	?
RSG-GAS (Indonesia)	1990	3.10^{14}/30 MW	?
Under Construction:			
FRM-II (Munich)		8.10^{14}/20 MW	~18, cost ~500M$[a]
Under Planning:			
TRRII (Taiwan)		~3.10^{14}/20 MW	cost ~300M$[a]
Australia		~3.10^{14}/20 MW	cost ~300M$[a]
IRF (Canada)		~3.10^{14}/20 MW	cost ~300M$[a]

[a] Construction; ?, information unknown.
Source: Presentation by John Rush of NIST to Panel in 8/97.

TABLE 3.4 US Spallation Sources

Facility	Agency	Year	Current Energy/ Target	Operation Cost ($ million FY 1996)
IPNS (ANL)	DOE	1981	6 kW/U	10
LANSCE (LANL)	DOE	1985	80 kW/W	11 ?
LANSCE upgrade	DOE	2002	6 new instruments, 160 kW	
Under Planning:				
SNS at ORNL			1-2 MW spallation source	cost ~1.3B$[a]

[a] Construction; ?, information unknown.
Source: Presentation by John Rush of NIST to Panel in 8/97.

ical systems (MEMS) exist at the National Center for Manufacturing Sciences (NCMS) and Sandia National Laboratory and are developing elsewhere, there is a growing demand for funding similar facilities for MEMS.

3.3.4 *Computing*

Computational research and engineering involving large-scale supercomputers and computer networks is of growing importance in solving materials problems at all scales from the subatomic to the macroscopic. Considerable progress has been made in developing models and simulations of complex materials phenomena based on first principles. Models and simulations are finding increasing use in supercomputer performance simulations of large-scale systems. Box 3.4 provides an example. However, important simulation and modeling research can be done even without a supercomputer.

TABLE 3.5 Spallation Sources Abroad

Facility	Year	Current Energy/Target	Operation Cost ($ million FY 1996)
ISIS (UK)	1985	160 kW	~25-30
KENS (Japan)	1980	3 kW	6
SINQ (Switzerland)	1996	800 kW (steady state)	?
Under Planning:			
N-arena (Japan)		600 kW (pulsed)	cost ~800 M$[a]
ESS (EEC, site to be determined)		~5 MW	cost 1B+$
AUSTRON (Austria)		~200 kW	cost ~400M$[a]

[a] Construction; ?, information unknown.
Source: Presentation by John Rush of NIST to Panel in 8/97.

TABLE 3.6 Synchrotron Light Source Operations in G7 Countries

G7 Country	Number of Synchrotron Light Rings	Total SR Operations, million $	Operations, Cost/GNP, x 10^6	Operations Cost per Capita, $
USA	9	183	27	0.73
Japan	9	126	27	1.01
Germany	6	55.5	29	0.68
France	3	41	31	0.71
Italy	2	32.6	29	0.57
UK	2	41.8	40	0.72
Canada	1	8.7	15	0.30

Source: Adapted from Table D-1 of the *Report of the Basic Energy Sciences Advisory Committee Panel on DOE Synchrotron Radiation Sources and Science*, DOE, November 1997.

NOTES:

- The **number of synchrotron light rigs** is the number of Synchrotron light source rings in operation, under construction, and expected to be approved. For European countries contributing to the European Synchrotron Light Source (ESRF) in Grenoble, France, ESRF is counted once for each country (Germany, France, Italy, UK). Industrially operated rings are not included.
- **US Total** includes the four DOE-funded facilities (Advanced Light Source (ALS) at Lawrence Berkeley National Laboratory, Advanced Photon Source (APS) at Argonne National Lab, National Synchrotron Light Source (NSLS) at Brookhaven National Lab, and Stanford Synchrotron Radiation Laboratory (SSRL) at Stanford University), two NSF-funded facilities (University of Wisconsin Synchrotron Radiation Center (Aladdin) and Cornell High Energy Synchrotron Source (CHESS), the CAMD facility at Louisiana State University and the SURF facility at NIST.
- **Japan** total includes the Photon Factory in Tsukuba, Spring-8 in Kamigori, and UVSOR in Okazaki. No operations costs were available for facilities at the Electrotechnical Laboratory (INIJI II, NIJI IV, and Teras) in Tsukuba, HISOR in Hiroshima, and Suburu in Himeji. Costs for industrially operated facilities (Sumitomo, NTT, Mitsubishi, IHH) not included.
- **Germany** total includes DELTA at the University of Dortmund and HASYLAB at Hamburg, projected costs of ANKA at Karlsruhe (estimated to be operational in 2000) and BESSY II in Berlin (estimated to be operational in 1999) and the ESRF contribution. It does not include expenses for BESSY I in Berlin which will be replaced by BESSY II. It also does not include the synchrotron at the University of Bonn. HASYLAB is an integral Part of the DESY laboratory, which is primarily a high-energy physics laboratory; the HASYLAB costs were estimated as a percentage of total DESY operations.
- **France** total includes operations/improvement expenses for the LURE source at Orsay and the ESRF contribution. It does not include expenses for the Soleil source (projected to become operational in 2003 but site not yet determined), which will largely, but not completely, replace LURE. The Soleil operations costs are expected to exceed those of LURE somewhat.
- **Italy** total includes Sincrotrone Trieste (ELETTRA) and the ESRF contribution. Costs for the Synchrotron radiation operations at the Daphne facility at the Frascati laboratory in Rome, from whom no data were received, are not included.
- **UK** total includes the Synchrotron Radiation Source (SRS) at the Daresburg Laboratory and the ESRF contribution.
- **Canada** total includes projected costs of the Canadian Light Source (CLS) at Saskatoon, Saskatchewan. Additional costs for Canadian beamlines at the University of Wisconsin Synchrotron Center (Aladdin) at Stoughton and the APS at Argonne are not included.

BOX 3.3 The MOSIS Service

The Metal-Oxide Semiconductor Implementation Service (MOSIS), based out of the Jet Propulsion Laboratory, is a low-cost prototyping and small-volume production service for custom and semicustom (VLSI) circuit development. Since its start, MOSIS has processed more than 30,000 integrated-circuit designs through several fabricators and technologies.

MOSIS collects designs from different sources, so individual designers can purchase small quantities by sharing the cost of fabrication. Instead of paying more than $50,000 for a dedicated run, users can get four packaged parts for a few hundred dollars. These prices dramatically lower the risk of VLSI prototyping.

Third- and fourth-year university undergraduates, and sometimes first-year graduate students, can design circuits to send to a MOSIS organization, which makes the "chips" and sends them back. The small businesses that use MOSIS have complete design security, even though the chip is shared.

By subcontracting for fabrication, MOSIS provides designers with a single interface to the US semiconductor industry, an industry known for its variety of interfaces and rapid technological changes. Using MOSIS drastically reduces the risk, time and cost of system development based on custom and semi-custom integrated circuits.

Source: Personal communication, B. Crowe, Defense Advanced Research Project Agency, November 1997.

3.3.5 Smaller Scale Facilities

Large facilities are an important leadership determinant, but the availability of smaller scale facilities also is critical, especially when they can be located near top researchers in that field. Some examples of small-scale facilities are x-ray characterization, surface analytic (ESCA, Auger), scanning probe instruments (STM, STS), crystal growth, and optical characterization. Many international awards, including Nobel prizes in physics and chemistry have been earned by researchers who used small-scale equipment. The plight of research universities in maintaining their facilities and instrumentation has been well documented. There continues to be concern among top university researchers that facilities and equipment for materials research in several foreign universities now outclass those at most universities in the United States. Of particular concern is the need for modern equipment for materials synthesis and processing, where the United States is lagging behind Europe and Japan.

BOX 3.4 A Cure for Composites

Manufacturing an airplane wing or fuselage is expensive, requiring materials, time, and energy to produce. If a wing or fuselage is flawed, it can exact a high cost. A group at the National Center for Supercomputing Applications (NCSA) at the University of Illinois, Urbana-Champaign, is attempting to reduce costly mistakes by improving one step in most manufacturing processes—composite curing.

The curing process is like a baseball game: It can be dull until there is a home run. The bulk of the curing chemical reaction occurs in a relatively short period (146–150 minutes), during which the temperature jumps sharply and so does the degree of cure.

Researchers at NCSA use a supercomputer to simulate a curing process called thermosetting to detect causes of common weaknesses, such as delaminations, in fiber–epoxy composites. If they know what causes structural failures in these materials, they can determine how to modify manufacturing to eliminate weaknesses. Composites are put into service in flight vehicles, automobiles, boats, pipelines, buildings, roads, bridges, and dozens of other products.

In addition to experimenting with manufacturing and curing processes, researchers are finding ways to improve other qualities of composites to make them strong, lightweight, durable, and inexpensive to produce.

Successfully manufacturing a composite requires the correct combination of temperature, pressure, and curing time. To find the best process for each material and service condition, several curing processes are tried and the materials are tested.

Curing tests can be simulated on supercomputers in relatively less time than is needed for laboratory tests. Simulations also save on the costly consumption of materials in the laboratory. Physical tests must still be performed, but less often, and their role is trimmed to determining material properties and validating computer simulations.

Aside from saving time and money, the simulations allow scientists to use helpful but complex mathematical methods, such as finite-element analysis, that would be almost impossible to use efficiently without parallel processing. Scientists use parallel processing to calculate temperatures, pressures, and displacements in thousands of composite sites simultaneously. Also, the simulations can provide scientists with information, about such things as a composite's internal stresses, that is difficult to obtain from actual tests.

In the future, composites will be simulated by virtual reality—an interactive, three-dimensional form of visualization—so that researchers can manipulate the manufacturing environment for optimum results. These simulations will offer the same advantages as those currently in use, but they will offer the advantage of an up-close, hands-on experience.

Source: Adapted from *A Cure for Composites* by Angela Bottum, supercomputing center website, University of Illinois. http://access.ncsa.uiuc.edu/Features/Composites/Composite.htm

3.4 Centers

Centers have become a key mechanism for supporting materials science and engineering research as they bring together researchers from many disciplines in one location. Several are supported by NSF:

- Materials Research Science and Engineering Centers (MRSECs),
- Science and Technology Centers (STCs),
- Engineering Research Centers (ERCs), and
- Institute for Mechanics and Materials (IMM).

MRSECs support interdisciplinary and multidisciplinary materials research and education. These centers have strong links to industry and other sectors and their goal is to establish a national network for university-based research. There are several major research areas:

- Surfaces (dynamics, reactions, catalysis);
- Structural materials, interfaces, grain boundaries, nano-mechanics;
- Polymeric materials, polymer science;
- Electronic and optical–photonic materials;
- Superconductivity, low temperature phenomena;
- Magnetic materials and structures;
- Nanophase and nanostructured materials, mesoscopic systems;
- Phases, phase transformations, order–disorder;
- Biomolecular materials, self-assembly, colloids;
- Advanced computation, modeling, materials theory; and
- Materials design synthesis and processing.

Eleven MRSECs were established in 1994; 13 more were established in 1996.

In addition to direct funding of materials research, universities can enter an NSF-wide competition to establish STCs and ERCs. Past COSEPUP (1996) and NAE (1989) reports have evaluated these activities. These programs are still soliciting new proposals.

Several STCs that focus on materials science and engineering research:

- Center for Quantized Electronic Structures (Quest),
- Center for Superconductivity,
- Advanced Liquid Crystalline Optical Materials (ALCOM), and
- Center for High Performance Polymeric Adhesives and Composites.

Current ERCs include

- Center for Particle Science and Technology,
- Center for Interfacial Engineering,

- Center for Advanced Electronic Materials Processing, and
- Center for Plasma Aided Manufacturing.

IMM was established in 1992 by NSF to promote interaction between the mechanics and materials communities by fostering activities of industrial relevance. The institute's office is at the University of California-San Diego but its activities and resources are scattered around the nation and the world. IMM's activities include workshops, short courses, and summer schools; scientific visits; planning meetings; and outreach and educational programs focused on a different theme each year. Recent themes have included scale-dependent mechanical phenomena in materials; aging, deterioration, and accelerated testing of materials; and material behavior in product and structural design.

All of these centers have strong multidisciplinary components, and they aid essential university–university and university–industry interactions at the doctoral and postdoctoral level.

3.5 Human Resources

As materials have taken on a more critical role in communications, transportation, and weapons, for example, the need for highly trained scientists has increased. Figure 3.1 shows the number of doctoral degrees awarded by US institutions for materials science and engineering from 1986 to 1995. However, as noted earlier, although materials science and engineering has grown as a distinct academic discipline, many researchers in the field have degrees in other areas. The chart shows the number of doctorates awarded overall and those awarded to

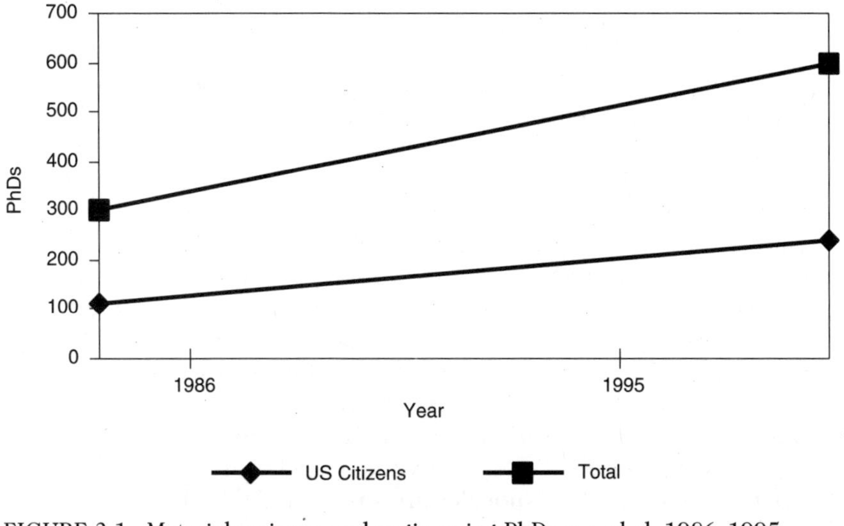

FIGURE 3.1 Materials science and engineering PhDs awarded, 1986–1995.

US citizens. The percentage going to citizens has remained at 40% for the period. Much has been written about dependence on noncitizens and whether it should be of concern. With the growth of high-tech education and industry in the Asian countries from which many of these noncitizens come, it seems probable that opportunities will expand for their return to their own countries, thereby reducing the net supply of trained scientists in the United States.

Tables 3.7 and 3.8 provide information on employment and occupational status, respectively, of PhDs in materials science in the United States. Figure 3.2 shows employment status. More than 60% of graduates consistently go to work in industry.

ASM-International has compiled a listing of materials faculties (ceramics, materials, metallurgy, and polymers) at educational institutions around the world (except for those in the former Soviet republics). Of these faculties, 169 are in North America (126 in the United States, 32 in Mexico, 11 in Canada), 117 are in Western Europe (40 in France, 28 in Great Britain), 27 are in Eastern Europe (not including the former Soviet republics), 84 are in Asia (29 in Japan, 19 in China), and 21 are in South America (dominated by Brazil with 20). The 85 Accreditation Board for Engineering and Technology (ABET) accredited undergraduate programs in the United States are divided into three main groups. Fifty are in the materials engineering group (28 of these are designated materials science and engineering), 22 are in the metallurgy group, and 11 are in the ceramic engineering group. One university has an accredited undergraduate program in welding engineering and there is 1 in plastics engineering. Two-thirds of the materials-related faculties at US institutions offer undergraduate degree programs.

The figures and tables presented here do not tell the complete story about graduates in materials-related programs, however. At the graduate level, some materials research and development programs occur in joint departments; others are department nonspecific. Part of the reason for this is the increasing role of computation in materials science and engineering, which often is carried out in other departments. Many PhD programs with a strong focus on materials are found in departments where materials is not the sole focus. This is particularly true for polymers, biomaterials, composites, catalysts, and electronic materials, among others. For example, there are 16 programs in polymer science in the United States, but most graduate students in polymers study in departments of chemistry and chemical engineering. Electrical engineering departments produce most of the graduates in semiconductor, magnetic, and optical–photonic materials. This makes an accurate accounting of the size and trends in graduate education difficult, but it does provide a picture of the diversity that enriches the field.

Data on women and minorities are shown in Table 3.9 and in Figures 3.3–3.5. Table 3.7 shows that the percentage of women PhDs in

TABLE 3.7 US Employment Status of Materials Science and Engineering PhDs, 1985-1995

	Survey Year					
	1985	1987	1989	1991	1993	1995
Tenured and tenure track faculty	92	95	249	209	337	426
Tenured Faculty	46	52	146	89	173	248
Tenured Track Faculty	46	43	103	120	164	178
Other Academic Positions	30	33	40	134	198	269
Postdoc Appointments—Academic	2	33	44	17	120	166
2 Year College Faculty	0	0	0	0	0	0
Industry	944	1495	1443	1946	2073	2592
Fed & Other Gvt Positions	23	38	85	121	171	234
Self Employed & Others	115	10	22	97	95	192
Postdoc Appointments—Other	67	9	11	0	88	82
Unemployed & Seeking	0	4	27	34	35	90
Elementary & High School Teachers	0	0	0	0	0	0
TOTAL	1273	1717	1921	2558	3117	4051
Tenured and Tenure Track Faculty	7.2%	5.5%	13.0%	8.2%	10.8%	10.5%
Tenured Faculty	3.6%	3.0%	7.6%	3.5%	5.6%	6.1%
Tenured Track Faculty	3.6%	2.5%	5.4%	4.7%	5.3%	4.4%
Other Academic Positions	2.4%	1.9%	2.1%	5.2%	6.4%	6.6%
Postdoc Appointments—Academic	0.2%	1.9%	2.3%	0.7%	3.8%	4.1%
2 Year College Faculty	0.0%	0.0%	0.0%	0.0%	0.0%	0.0%
Industry	74.2%	87.1%	75.1%	76.1%	66.5%	64.0%
Fed & Other Gvt Positions	1.8%	2.2%	4.4%	4.7%	5.5%	5.8%
Self Employed & Others	9.0%	0.6%	1.1%	3.8%	3.0%	4.7%
Postdoc Appointments—Other	5.3%	0.5%	0.6%	0.0%	2.8%	2.0%
Unemployed & Seeking	0.0%	0.2%	1.4%	1.3%	1.1%	2.2%
Elementary and High School Teachers	0.0%	0.0%	0.0%	0.0%	0.0%	0.0%

Source: Analysis conducted by the National Research Council's Office of Scientific and Engineering Personnel of data from Survey of Doctorate Recipients (SDR) for this study.

materials-related subfields in the physical sciences is greater (17.4%) than is the percentage of women PhDs in materials-related subfields in engineering (15.9%). As a percentage of the total number of women PhDs in the physical sciences and engineering, the percentage of PhDs in materials-related subfields in the physical sciences (6.1%) is lower than is the corresponding percentage (15.3%) in engineering.

Table 3.10 shows that, among the G-7 nations,

- Italy ranks first for percentage of first degrees to women in the natural sciences;

TABLE 3.8 US Occupational Status of Materials Science and Engineering PhDs, 1985-1995

	Survey Year					
	1985	1987	1989	1991	1993	1995
Total Research	857	1062	1200	1507	2127	2379
Basic Research	79	93	206	186	245	229
Applied Research	713	801	793	914	1209	1244
Development	65	168	201	407	673	906
Research Management	251	250	244	349	N/A	N/A
Management Other	26	0	4	29	N/A	N/A
Management	N/A	N/A	N/A	N/A	351	501
Teaching	10	39	79	90	187	284
Professional Services	0	0	0	0	5	26
Consulting	2	10	0	98	N/A	N/A
Computing	N/A	N/A	N/A	6	18	111
Other Work Activities/No Response	127	352	367	445	394	660
Federal Support	521	777	994	1107	970	1436
No Federal Support/No Response	752	936	900	1417	2112	2525
Total	1273	1713	1894	2524	3082	3961
Total Research	67.3%	62.0%	63.4%	59.7%	69.0%	60.1%
Basic Research	6.2%	5.4%	10.9%	7.4%	7.9%	5.8%
Applied Research	56.0%	46.8%	41.9%	36.2%	39.2%	31.4%
Development	5.1%	9.8%	10.6%	16.1%	21.8%	22.9%
Research Management	19.7%	14.6%	12.9%	13.8%	N/A	N/A
Management Other	2.0%	0.0%	0.2%	1.1%	N/A	N/A
Management	N/A	N/A	N/A	N/A	11.4%	12.6%
Teaching	0.8%	2.3%	4.2%	3.6%	6.1%	7.2%
Professional Services	0.0%	0.0%	0.0%	0.0%	0.2%	0.7%
Consulting	0.2%	0.6%	0.0%	3.9%	N/A	N/A
Computing	N/A	N/A	N/A	0.2%	0.6%	2.8%
Other Work Activities/No Response	10.0%	20.5%	19.4%	17.6%	12.8%	16.7%
Federal Support	40.9%	45.4%	52.5%	43.9%	31.5%	36.3%
No Federal Support/No Response	59.1%	54.6%	47.5%	56.1%	68.5%	63.7%

Source: Analysis conducted by the National Research Council's Office of Scientific and Engineering Personnel of data from Survey of Doctorate Recipients (SDR) for this study.

- France ranks first for percentage of first degrees to women in engineering;
- The United States ranks third for percentage of first degrees to women in the natural sciences as well as in the percentage of first degrees to women in engineering; and
- Japan ranks last in percentage of first degrees to women in the natural sciences and engineering.

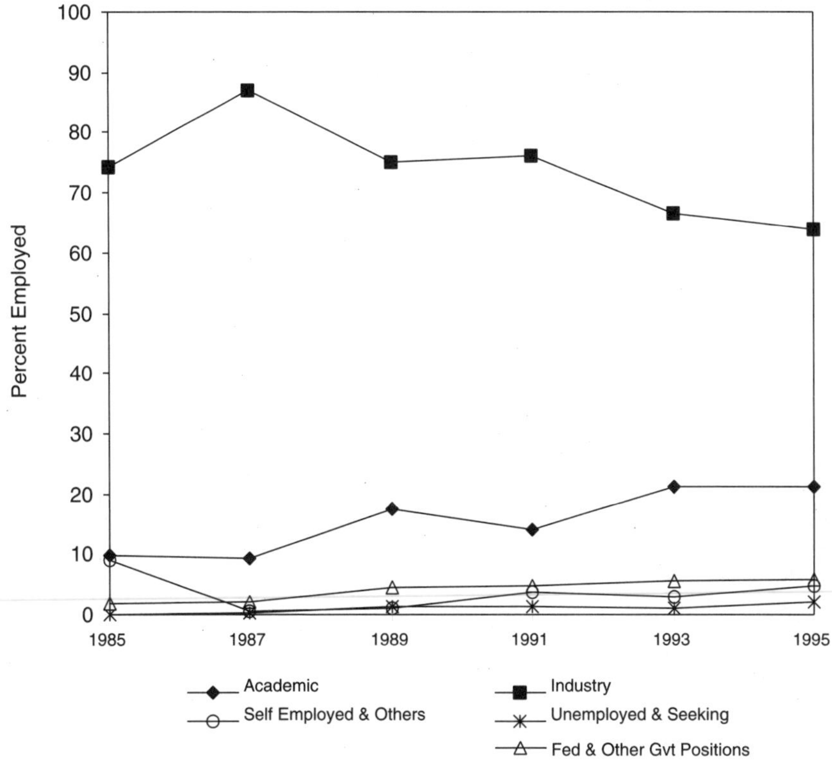

FIGURE 3.2 Employment status of PhD materials scientists, 1985. Source: Analysis conducted by the National Research Council's Office of Scientific and Engineering Personnel.

As shown in Figure 3.3, metallurgical–materials engineering has a significantly lower representation of black, non-Hispanic graduate students than do all-engineering (Figure 3.4) and the sciences (Figure 3.5). Metallurgical–materials engineering has a substantially greater representation of noncitizens among graduate students than do all-engineering and the sciences. The percentage of foreign students is only 7.1% lower than that of white, non-Hispanic graduate students.

3.6 Funding

Obviously, adequate funding is a necessary element of any research effort. Because the US innovation system has many funding sources and because materials science is a diverse field, it is difficult to determine the amount of funding, much less judge its adequacy. During the Bush Administration, the Federal Coordinating Council for Science, Engineering, and Technology (FCCSET) undertook a broad analysis of federal

TABLE 3.9 Number of Doctorate Recipients by Gender and Subfield

	Number			Percentage	
Field and Subfield	Total	Men	Women	Men	Women
Physical Sciences	6,806	5,307	1,499	78.0	22.0
• Physics					
Polymer	23	20	3	87.0	13.0
Solid-state and low temperature	371	314	57	84.6	15.4
• Chemistry					
Polymer	116	88	28	75.9	24.1
• Earth, atmosphere, and marine					
Minerology and petrology	19	15	4	78.9	21.1
Totals	529	437	92	82.6	17.4
Percentage of physical sciences	7.8	8.2	6.1		
Engineering	6,007	5,313	694	88.4	11.6
• Ceramic sciences	39	34	5	87.2	12.8
• Materials sciences	476	392	84	82.4	17.6
• Metallurgical	73	68	5	93.2	6.8
• Mining and mineral	19	19	0	100	0
• Polymer and plastics	58	46	12	79.3	20.7
Totals	665	559	106	84.1	15.9
Percentage of engineering	11.1	10.5	15.3		

Source: Appendix A, Table A-1 from the NRC Survey of Earned Doctorates.

support in several research areas, one of which was materials. The results for all agencies for fiscal years 1992, 1993, and 1994 are shown in Figure 3.6 and Table 3.11. Updated information for NSF's Division of Materials Research is shown in Figure 3.7. The average annualized award size for that division compared with others within NSF's mathematics and physical sciences directorate is shown in Figure 3.8. At

TABLE 3.10 Percentage of First Degrees in Science and Engineering to Women, G-6 Nations

Country	Natural Sciences	Engineering
Italy	54	8
United Kingdom	44	26
United States	42	16
Germany	40	11
France	35	19
Japan	19	4

Source: Table derived from S&E Indicators, Appendix Table 2-5.

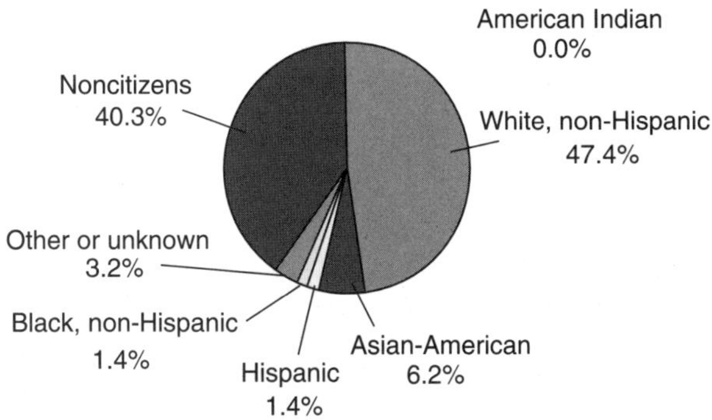

FIGURE 3.3 Metallurgical-materials engineering graduate students in all institutions, by race-ethnicity and citizenship, 1993. Source: National Science Foundation, Women, Minorities, and Persons with Disabilities in Science and Engineering: 1996, NSF 96-311.

about $125,000 each, awards for materials research were exceeded only by physics awards at about $170,000 each. Figure 3.9 shows the budget for permanent equipment, which in 1996, was about $18 million. Approximately 20% of the fiscal year 1998 request from NSF was for facilities.

Materials research and development is defined broadly here to include scientific and engineering research on substances in any form and at any stage of preparation, fabrication, manufacture, recycle, or

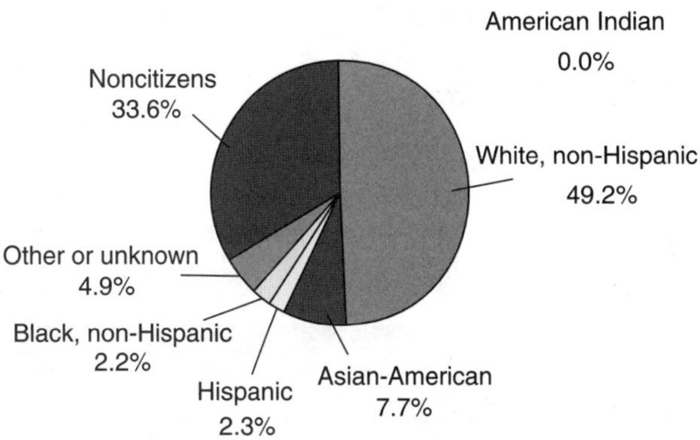

FIGURE 3.4 All engineering graduate students in all institutions, by race-ethnicity and citizenship, 1993. Source: National Science Foundation, Women, Minorities, and Persons with Disabilities in Science and Engineering: 1996, NSF 96-311.

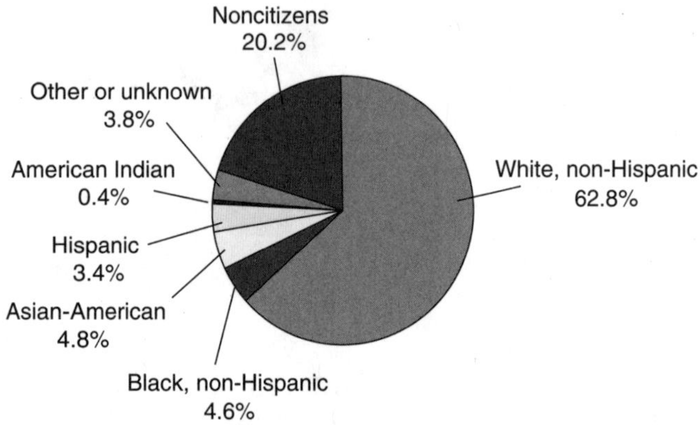

FIGURE 3.5 All science graduate students in all institutions, by race-ethnicity and citizenship, 1993. Source: National Science Foundation, Women, Minorities, and Persons with Disabilities in Science and Engineering: 1996, NSF 96-311.

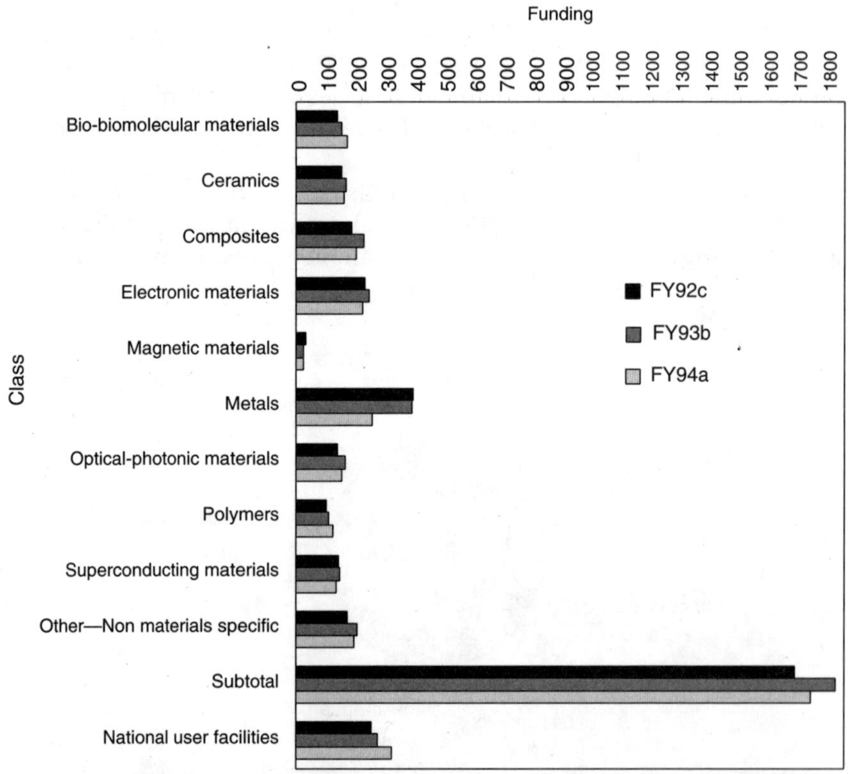

FIGURE 3.6 Federal R&D budget by materials class, in millions of US dollars. Excludes classified research and development and most development activities funded under DOD's specific systems R&D programs. a, President's budget request; b, congressional appropriations; c, actual expenditures

TABLE 3.11 Federal R&D Budget for Materials Research by Agency, in Millions of US Dollars

Agency	Year	Total Program
Department of Commerce	FY94[a]	56.7
	FY93[b]	48.4
	FY92[c]	42.6
Department of Defense[d]	FY94	421.7
	FY93	557.7
	FY92	530.9
Department of Energy[d]	FY94	941.5
	FY93	914.0
	FY92	862.5
Department of the Interior	FY94	21.5
	FY93	24.9
	FY92	25.2
Department of Transportation	FY94	12.7
	FY93	14.9
	FY92	11.0
Environmental Protection Agency	FY94	4.5
	FY93	4.5
	FY92	3.5
Health and Human Services	FY94	92.9
	FY93	85.9
	FY92	79.6
National Aeronautics and Space Administration	FY94	131.1
	FY93	102.8
	FY92	76.3
National Science Foundation	FY94	328.0
	FY93	303.6
	FY92	265.6
United States Department of Agriculture	FY94	45.8
	FY93	37.4
	FY92	36.3

Note: Total program includes construction and operating costs for major national user facilities.
[a] President's budget request.
[b] Congressional appropriations.
[c] Actual expenditures.
[d] Excludes classified research and development and most development activities funded under DOD's specific systems R&D programs.

disposal. Program types vary and include basic research and technology development, device or process development, and the gathering or analysis of data on various materials. The fiscal year 1994 budget contains significant activities in synthesis and processing and in theory, modeling, and simulation. These two components were emphasized in a major National Research Council (1989) study on materials science and

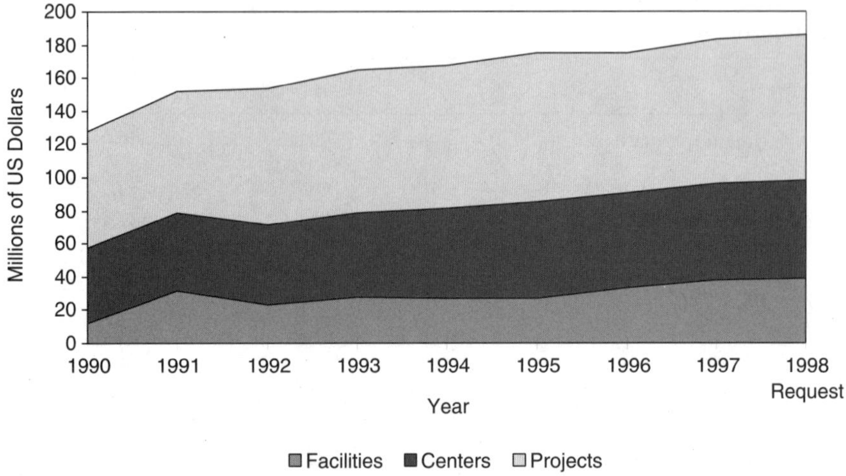

FIGURE 3.7 National Science Foundation Division of Materials Research Budget, 1990–1998, in millions of US dollars. Source: http://www.nsf.gov/mps/dmr/chart1.htm with updated information.

engineering that incorporated considerable input from industry. The FCCSET study concludes that the budget provided adequate funding to achieve breakthroughs in energy, environment, health and safety, information and communication, infrastructure, national security, and transportation. However, this conclusion was reached before more

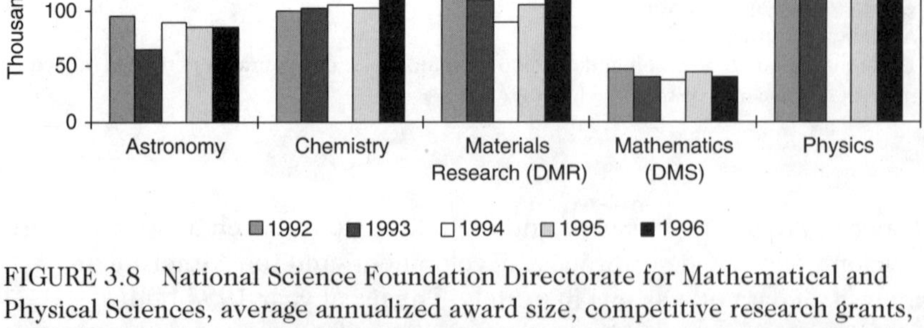

FIGURE 3.8 National Science Foundation Directorate for Mathematical and Physical Sciences, average annualized award size, competitive research grants, 1992–1996, in thousands of US dollars. Includes instrumentation awards and programs such as PFF, NYVPYI, and CARCCR. Source: http://www.nsf.gov/mps/portfoli/maverage.htm.

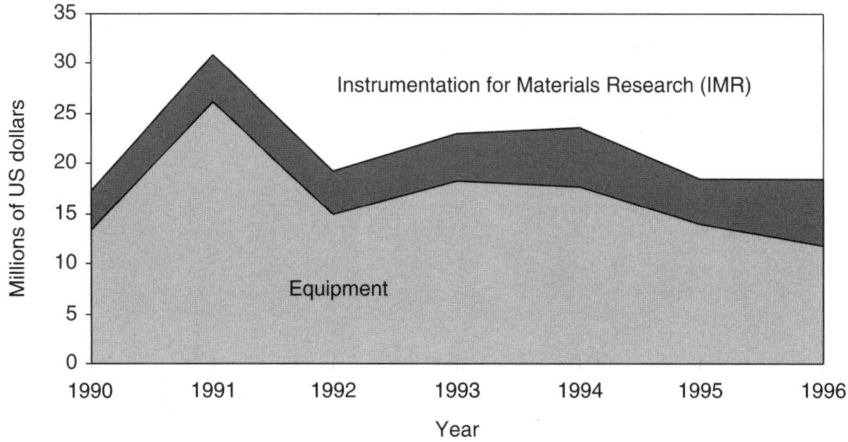

FIGURE 3.9 National Science Foundation Division of Materials Research, permanent equipment budget, 1990–1996, in millions of US dollars. Source: http://www.nsf.gov/mps/dmr/chart1.htm with updated information.

recent drawdowns in federal investment in research and development (OSTP 1995).

The percentage of research and development investment for materials and industrial technology within the European Union increased from 11% in 1984 to 16% in 1987 and has stayed constant. Materials technologies constituted 9.1% of Eureka projects in 1992. Eureka is a cooperative effort in research and development (outside the European Union program) geared toward industrial and applied research and development (NSF 1996).

In general, it is important to note that several successful nations in high-tech industries (e.g. Japan) follow a different paradigm and strategy in allocating research resources.

4

BENCHMARKING RESULTS

4.1 Approach

The approach taken by the Panel to assess the strength of the field of materials research in the United States relative to other countries was as follows:

- By communicating with colleagues in the United States and abroad, panel members scripted the content of a fictitious international conference covering the nine subfields of materials science and engineering. Panel members asked colleagues to identify 5 or 6 "hot topics" in each subfield and 8 to 10 of the very best people in the world working in each topical area.
- The top 3–4 awards and prizes given in each area and their recipients for the past five years were identified.
- The most significant advances in materials science and engineering research of the past five years were identified.
- Assessments of the state of research in each topical area in the United States compared with that of other nations were solicited.
- The leading journals or other periodicals that could be used as references for the assessment were identified.

The information was used to construct tables that characterize the relative position of the United States in each of 9 subfields now and in the future (Appendix B). The first half of each table ranks the current US position relative to the world materials community for each subfield. A scoring system, with 1 representing "forefront", 3 representing "among world leaders", and 5 representing "behind world leaders", was used. The second half of each table is an assessment of the likely future position of the United States relative to the world materials community. Here 1 represents "gaining or extending", 3 represents "maintaining", and 5 represents "losing". Although the conference approach does not constitute a systematic assessment and is somewhat subjective, it is the same approach leaders of the field would use to organize world conferences to feature the "best of the best." And in conducting this analysis, panel members relied not only on their own judgment but that of their colleagues.

A report from the United Kingdom (UK 1997) compared by field and country the number of total publications and the relative citation impact. The top 5 countries for the materials science field by total citations are:

1. United States
2. Japan
3. Germany
4. United Kingdom
5. France

The measure used in the United Kingdom report, "relative citation impact", is the country's share of the world's citations in the field, divided by its share of world publications in the field. It can be thought of as a comparison of a country's citation rate for a particular field with the world's citation rate for the field. A relative citation impact greater than 1 shows that the country's rate for the field is higher than the world's. According to the report, it is a measure of both the influence and the visibility of a country's research (as disseminated through publications) and it gives some indication of the quality of the average paper. The top 5 countries by relative citation index are:

6. United States
7. Denmark
8. Netherlands
9. Israel
10. Switzerland

4.2 Assessment of Current Leadership

US research in materials science and engineering is strong, based on assessment of the subfields. Across subfields, US researchers are among the world leaders. Within subfields there are a few topical areas in which the United States is at the forefront and a few topical areas where the US has little presence or is behind the world leaders. (A general example of the latter is in materials synthesis and processing, where greater scientific emphasis is needed here. Europe is probably ahead in synthesis, and Japan leads in processing. If US universities were to focus research more on "green" processing for sustainable development, US industry could benefit.) For most subfields and topics, however, the United States ranks among world leaders. Comments on the analysis of each of the subfields are found below.

4.2.1 Biomaterials

The subfield of biomaterials involves at least four classes: synthetic or modified natural materials used in medicine and biology; natural materials or artificial materials that emulate natural materials; "smart" materials, such as those that respond to a specific stimulus; and hybrid materials consisting of synthetic and living cellular components.

Much of the research in biomaterials originated in the United States, and the country still maintains an intellectual lead in most areas, particularly basic research. This is suggested by the number of US keynote and plenary speakers at international conferences and by the number of US scientists who receive major awards in the field. The biomaterials industry contributes a positive balance of payments for the United States. The applications of biomaterials also helps the US economy by reducing the cost of health care.

The factors that influence US performance in biomaterials include a strong basic research organization that offers graduate and postdoctoral students opportunities to study, the ethnic diversity in our graduate and postdoctoral student populations, strong journals, a strong medical device industry, venture capital–commercialization potential, and the sheer number of researchers in the field.

The current hot topics are in tissue engineering, protein analogues, molecular architecture, biomimetics (bone-like material), contemporary diagnostic systems, advanced controlled-release systems, and bone materials. The United States is currently at the forefront in tissue engineering, protein analogues, and advanced controlled-release systems. It is among the world leaders in all other areas. The United States is seeing strong competition from Germany and Japan in molecular architecture. There are some areas of biomaterials research in which the United States no longer is the clear leader: hydrogels, proteins at interfaces, artificial hearts, surface molecular engineering sensors, and diagnostics.

4.2.2 Ceramics

Most consumers think of "ceramics" as freezer-to-oven cookware—this wide application for glass ceramics is possible because of the materials' low expansion coefficient and ability to withstand a range of temperatures without cracking.

There have been only a few applications for tailored ceramics based on their thermostructural and electromechanical properties because of their relatively high cost and underdeveloped design and reliability protocols. The most significant application has been as electronic substrates (mostly alumina), as well as some commercialization of SiC and Si_3N_4. The latter are being used as wear components in the industrial sector and as valves and bearings primarily in aerospace applications. Further implementation is expected to be controlled primarily by the success of initiatives that reduce the manufacturing cost. Although research funding in this area is declining worldwide, new areas of activity are providing important opportunities for basic research that should be exploited if the United States is to remain in the forefront. Four of these are as follows:

- The use of Microeclectromechanical systems (MEMS) based mini-heat engines, made from SiC, as small high-power-density energy sources constitute an exciting initiative. For example, gas turbine engines, rocket engines, and coolers about a centimeter in diameter and a few millimeters thick that should produce power and pump heat in the 10–100 W range could be made using MEMS technology. Ceramics appear to be perfect for this implementation because the small component size mitigates the weakest-link nature of ceramics' mechanical strength.
- Recently discovered single-crystal ferroelectrics offer a completely new range of options as actuators and capacitors and in robotics.
- High thermal conductivity, low-expansion dielectrics, such as AlN, needed for heat dissipation in power electronics, are being developed by industry with minimal basic research.
- High-performance films and coatings for thermal protection (ZrO_2 alloys), lubrication and durability (diamondlike coatings, TiN), and implants (hydroxyapetite), among other uses, are being exploited by US industry. An invigorated academic effort is needed in films and coatings.

The United States and Japan share leadership in ceramics: Japan dominates in manufacturing technology, and the United States leads in basic research. There are also strong research activities in Germany.

In the field of functional–electronic ceramics, the United States is among the world leaders in all areas other than integrated micromagnetics; Japan is the clear leader there. Current forefront areas include microwave dielectrics, sol-gel-derived materials, self-assembled materials, thin-film synthesis, three dimensional microporous silicates, multilayer ferrite processing, and integrated micromagnetics on silicon.

Although US research in ceramics is in a world leadership position, the companies that capitalize on that research are Japanese-owned.

This question is the subject of another COSEPUP study to be released later this year.

4.2.3 Composites

The United States has been a leader in the development of polymer, metal, and ceramic matrix composites over the past 30 years. Vigorous efforts in France and Japan have matched those here, in some cases. There also have been several important discoveries in the United Kingdom. The activity has mostly found application in aerospace and has been supported almost exclusively by the Department of Defense (DOD) and the National Aeronautics and Space Administration (NASA). More recently, there has been significant activity in the automotive and energy sectors, with Department of Energy (DOE) support. Federal support for this research at universities has essentially stopped, because of changing DOD, DOE, and NASA priorities. The remaining, relatively small academic efforts in academia are supported largely by industry, are very applied, and have short-term goals. Thus, continued leadership in this field is uncertain.

In polymer composites, a new approach to design and manufacturing that cuts costs has begun to replace the "black aluminum" approach. The "black aluminum" method, used for decades to design polymer composites, is a design protocol that treats composites as if they were metals, using the same types of data and the same design rules. The problem with this approach is that the real advantages that could accrue to using composites are negated at the outset. This has been done for decades because of conservatism. Only now are new design approaches that take advantage of the attributes of composites (and as anisotropy, integrated structures) being implemented. The new method uses low-temperature-curing polymers for matrices. Electron beams achieve homogeneous curing. This combination produces fewer distortions and enables the manufacturing of large integral components. There are significant opportunities for new basic research as the technology progresses in industry. These include the development of low-temperature-curing polymers with greater toughness and higher glass transition temperatures and design-and-testing protocols explicitly relevant to large integral components along with included subelements.

Ceramic matrix composites are important for the aerospace and energy sectors. Industry activity is appreciable; federal support comes primarily from NASA and DOE. There are exciting new developments in all-oxide composites that could survive high temperatures in aggressive environments. The basic research needed to support the development and use of these materials includes studies of oxide–oxide bonding, creep, internal friction, and delamination. The issues are addressed in the new National Materials Advisory Board report, *Advanced Fibers for High Temperature Ceramic Composites* (NRC 1997).

Major shifts in emphasis are occurring in research on metal matrix composites because of new applications in the aerospace, communications, electronics, and automotive industries. These applications are primarily based on aluminum alloys reinforced with SiC or Al_2O_3, as particles, whiskers, or fibers. These materials have become less expensive to produce, they are fairly rigid (an advantage when combined with acceptable toughness and ductility), and have high fatigue thresholds. They are useful in electronic applications because of their low thermal expansion coefficients and high thermal conductivity. The effort to develop in titanium matrix composites reinforced with SiC fibers has nearly stopped in the United States because of the great expense involved. Research continues in Japan, China, and the United Kingdom.

With the much-diminished DOD emphasis on structural materials, basic research on composites has decreased dramatically in the past 2 years. This is happening as new applications are emerging and new opportunities have arisen for basic research. The exception is in smart materials and systems. Here, the polymer composite is a host for sensors and actuators that enable shape changes to be accomplished with rapid response times. A major effort is continuing in this area in the United States.

Current hot topics in research include smart composites that incorporate sensors and actuators, polymer matrices for ambient curing (for example with electron beams), the manufacture of large integrated components, high-temperature oxide materials, the tailored use of preferred crystallographic textures, and particle-reinforced alloys. In this last area, there is an opportunity to extend the crack-arresting concepts that have been applied to metal matrix composites to other materials, particularly titanium alloys, and to elucidate at a fundamental level how improved fatigue performance can be achieved by reinforcement.

4.2.4 Magnetic Materials

Research on magnetism and magnetic materials, which in the US has been strongly influenced by applications, has declined since the 1970s. There were major breakthroughs in hard and soft magnetic materials (Allied developed soft amorphous materials and General Motors and Sumitomo developed neodymium–iron–boron permanent magnets), but markets were not large enough to support a large research community. In Europe and Japan, on the other hand, basic research in dilute magnetic alloys and critical phenomena sustained an interest in magnetism and led to investment in the necessary support facilities, such as neutron sources and high magnetic fields. Two of the hot areas of recording today, giant magnetoresistance and spin-dependent tunneling, were discovered in France.

In the 1980s a "killer application" developed in the United States—digital magnetic recording for computer systems. This is now more than a $50 billion industry. The technology was developed mainly at IBM, and US companies such as IBM, Seagate, and Quantum are the market leaders.

Although magnetic studies largely disappeared from US universities in the 1970s, there has been a concerted effort over the past 2 decades to rebuild interest. There are now several centers for magnetic materials study as part of engineering schools, and basic research efforts are small but growing. One measure of US participation in magnetic materials research is the annual conference on Magnetism and Magnetic Materials. Figure 4.1 shows that, since 1989, the United States has contributed a consistent and respectable percentage of papers to this conference.

Giant magnetoresistance is being used as the reading element in high-density recording heads through a structure known as a spin valve, which consists of two thin magnetic films with different coercivites separated by a very thin (20Å) conductor. The application of spin valve heads requires a fundamental understanding of the magnetic interactions within and between the films. Thus, the area of magnetic interfaces is also hot. How these materials behave with temperature and their corrosion resistance are active areas of research.

Large magnetoresistance effects have been observed in manganese-oxide perovskites. The effect is so large, it is called "colossal" magnetoresistance. Although the recording community does not believe these materials will compete with spin valves, their structural similarities to high-temperature superconductors has generated interest in the United States and abroad.

The recent discovery of giant magnetostriction in a new class of materials–shape memory alloys that are magnetic—promises to produce another hot area. Researchers at the University of Minnesota and the University of Maryland have observed magnetostriction of 1.2% in single crystal, prestressed NiMnGa. These materials could replace piezoelectrics in many applications where a more robust material and larger strain is required. They already appear superior to the rare-earth-containing materials such as Terfenol.

The current forefront topics in magnetic materials are nanostructures, colossal magnetoresistance, magnetic multilayers including magnetic properties of thin layers (first-principles calculations and micromagnetics), magnetic coupling between layers (anisotropic exchange and biquadratic exchange), and transport (giant magnetoresistance and tunneling). Disk storage has become the driver for research in the United States, and university centers are being developed. Although the United States is at the forefront in the device area, solid basic research is being done at centers elsewhere.

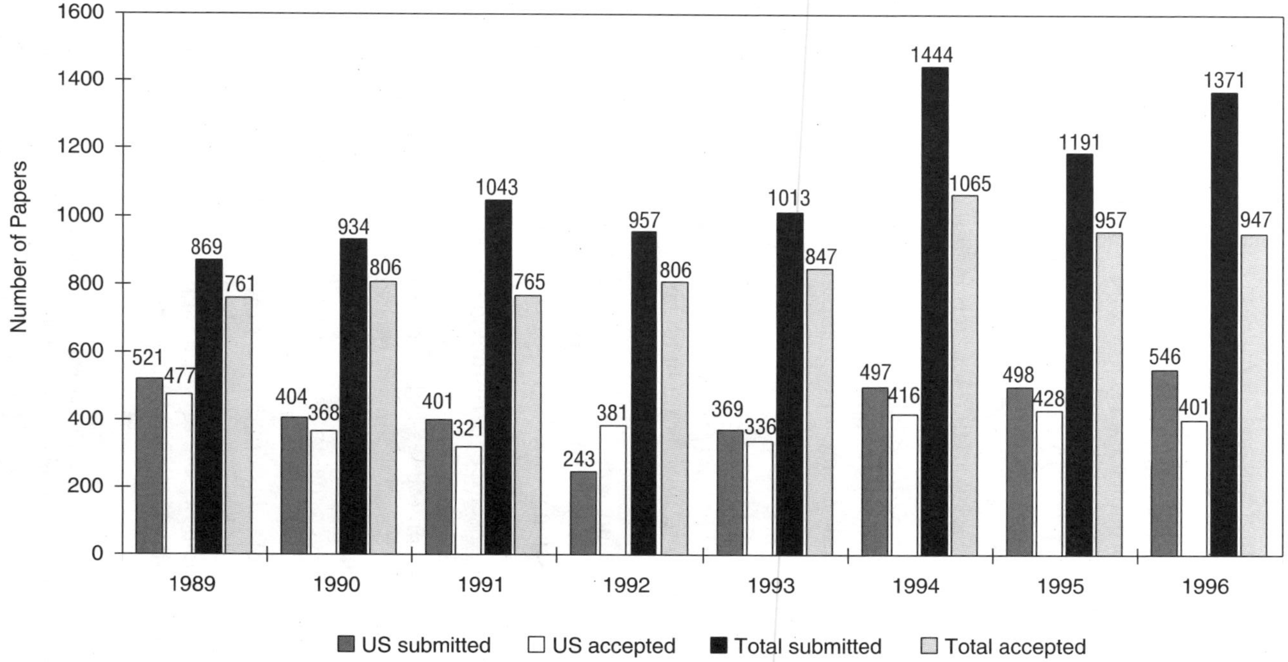

FIGURE 4.1 Papers submitted and accepted for Magnetism and Magnetic Materials Annual Conferences, 1989–1996. Source: Magnetism and Magnetic Materials Conference.

4.2.5 Metals

The performance of almost any product is limited by the materials of which it is made, and in many products, space vehicles, for example, the value of overcoming performance barriers is quite high. In automobiles, the value of improved performance is less well understood, but is currently motivated by energy efficiency targets while constrained by cost. These considerations become important as more of the drive for improved materials performance comes from industry sectors other than defense.

The cost and time required to develop and deploy new materials are a major problem regardless of the application or industry. Product development times have been and will continue to be dramatically shortened, but the time it takes to develop new materials development has not been reduced significantly, and this creates a barrier to achieving optimum final product performance. One key to solving the cost and time constraints for materials development is computational materials science and engineering. Modeling, simulation, and experimentation are used to study

- Materials synthesis;
- Microstructure evolution (precipitation, recrystallization, phase transformation, defect structures, grain boundaries);
- Plastic deformation behavior;
- Materials performance and properties (strength, ductility, fracture toughness, formability, fatigue, corrosion/durability);
- Complex processing methods (chemical, thermal, and mechanical processing of materials)*; and
- Component and assembly performance.

This list of topics is sometimes called the structure–process–product continuum. The research aims at quantitative explanations of the relationships between processing and structure, between structure and properties, and between materials properties and product performance. Yet another way to describe this is the integration of dimensional scales—from atomic clusters to final products—and the integration of models of materials behavior, materials processing, and materials product performance to allow concurrent design of products, materials (composition and structure), and manufacturing processes.

*Complex processing methods include net shape processes such as isothermal–superplastic forming and computer-aided precision casting and machining, where laser and electron beams often are used as cutting tools. Net shape isothermal–superplastic forming is the most typical example of "complex processing." Recently, various high-temperature components, such as turbine disks have been produced from nickel-based superalloy powders by net shape isothermal–superplastic forming.

The current hot topics in this subfield are surface treatments, net shape processes, intermetallics and other high-temperature alloys, theory and modeling, magnetoresistance, hydrogen-absorbing materials, bulk amorphous materials, quasicrystalline materials, nanostructured materials, and cellular metals. The United States is especially strong in intermetallics, theory and modeling, and advanced processing of metallic alloys to net shape.

The United States is in a leadership position in most of these subfields; however, there is significant capability in the United Kingdom, Germany, France, and Japan. There are also significant resources in Eastern Europe and Russia, but they are not necessarily well-funded and they are not as well known. The United States is lagging in research on hydrogen-absorbing materials; Germany and Japan have a strong position. Battery development for applications in computers and motor vehicles is a pull for electrode research and development. Bulk glass-forming alloys have been studied in the United States, but intensive work also is going on in Japan. The United States is at the forefront of theory and modeling of metals. The topical areas of current activity in theory and modeling include atomic bonding, crystal structure, microstructure evolution phase diagrams, and phase transformations. Good work is going on in Europe, Japan, and the United States on quantitative explanations and modeling of the plastic deformation of metals.

4.2.6 Electronic and Optical-Photonic Materials

Electronic materials encompass a broad field, with semiconductors obviously at the center. However, metals, dielectrics, ceramics and polymers also are in this group. In terms of functional applications, which are process intensive, these materials are found in even more diverse areas: lithography, interconnect, packaging, display and storage materials. Generally the United States is the world leader in most of these subfields and a close competitor in others.

In the display subfield, the United States is behind Japan in liquid crystals and wide-gap III-V compounds (GaN). In liquid crystals, the United States is unlikely to catch up to Japan. In the case of flat-panel displays, for example, because US industry has less than 10% of the market (most displays come from Japan with a growing number from Korea and Taiwan), future industrial research and development is likely to be limited. In wide-gap III-V compounds, the United States has made a major investment to try to close this lead, and the lead in Japan is expected to diminish over time.

Semiconductors have been an important subject of US materials science and engineering since the invention of the transistor and the subsequent progress made in the quality and purity of silicon and germanium. The US also led in developing compound semiconductors (GaAs) for optical, electronic and communication applications, although these materials were first formally recognized in Germany.* As the synthesizing and processing technologies of basic materials matured, improvements were incremental and manufacturers profit margins declined. Materials suppliers consolidated and they have moved offshore, mostly to Far East Asia.

With the approach of the anticipated limit for conventional technologies in the silicon-based electronics industry, great emphasis was placed on clever design and processing and on advanced lithography. Attention also was paid to more innovative considerations in the use of silicon on insulators, silicon–germanium alloys, low-loss dielectrics, and high-conductivity metal interconnects. Again, the United States and Japan lead the field in these areas.

One exciting development in recent years is in the area of nanostructures. This started with the epitaxial and vapor phase growth of semiconductor quantum layers and now includes many material systems: self-organized particles, clusters, and tubes; caged ensembles; porous solids and surfaces; and colloidal crystallites. With feature sizes of just nanometers, these materials exhibit unusual electronic and optical properties that offer enormous opportunities in materials research and technological applications. Most of the systems were developed in the United States, which has largely maintained its leadership despite worldwide growth in research. A word of caution is called for, however. Many of the central research laboratories of the large industrial corporations initiated and carried out studies in these subfields. The redirection from basic to applied research and technology, although accelerating the US innovation process, is shifting future exploratory research to universities.

Optical–photonics materials research will provide many opportunities for exploring the most fundamental aspects of materials. Examining optical phenomena, such as imaging, holographic storage, electro–optic and photorefractive effects, optical fiber nonlinearities, and complete modeling of optical–electronic integrated circuits are subjects of great interest. Just as past exploratory materials research uncovered soliton and other phenomena, the current basic research being devoted around the globe to the fabrication and theoretical modeling of photonic bandgap materials should yield exciting and unexpected results.

Wavelength-selectable, blue-green, all-solid-state, and microlasers are internationally hot topics. Recent breakthroughs in photonic

*III-V Compounds, such as GaAs, were first recognized in Germany as intermetallic semiconductors (Welker in 1952). Subsequent development was done worldwide, with the United States at the lead. The injection laser was developed by several US groups (Hall, Nathan, Quist and Holonyak, all in 1962). The heterojunction laser was developed by Russian researchers (for example, Alferov 1967).

bandgap and lattice engineering, and in atomic layer epitaxy, provide unprecedented possibilities for leaps in device and component research. These materials-processing methods also should contribute to even more rapid progress in electronic–photonic integration. Much also is expected from MEM research, which will continue to provide a powerful tool for miniaturization of mechanical systems.

Organic materials have become the subject of exciting optical–photonic research. Organic lasers, organic light-emitting diodes, all-organic transistors, and plastic fibers are receiving global attention within small, focused research groups and also from larger development organizations. The United States and others are equivalent in these areas. The United States has established a substantial lead in another area—the fabrication of complex, optically functional surfaces, components, and devices using elastomers as starting materials. Polymeric replicas, patterned with microstructures on their surfaces, are fabricated and used to construct lenses, mirrors diffraction gratings, and photothermal detectors. The clear lead of the United States in this new area is not expected to persist.

US competitiveness in photonics and microelectronics has been well served by the establishment of academic centers. Strong support for these facilities through the National Science Foundation is vital. US industrial–academic partnerships also are needed to advance the research required to win the global competition in optical networking.

The United States is currently among the leaders or at the forefront in optical–photonic and electronic materials. Forefront topics include wide direct-gap and wide indirect-gap semiconductors, wafer bonding, oxide confinement in vertical cavity surface emitting lasers, interconnects (copper and low-K dielectrics), engineered optical materials (periodically poled nonlinear optical materials and engineered organic and polymeric materials for nonlinear optics), nanostructured materials, magnetoresistive materials, holographic storage materials, photonic bandgap materials, and packaging.

More information is available in a recent report from the NRC Committee on Optical Science and Engineering (COSE) entitled *Harnessing Light: Optical Science and Engineering for the 21st Century* (NRC, 1998). This report discusses the state of the optics industry and of research and education in optics, and identifies actions that could enhance the field's contribution to society and facilitate its technical development.

4.2.7 Superconducting Materials

US scientists are in a strong leadership position in nearly all subtopical areas of superconducting materials, but they do not dominate in any. The United States has been at the forefront in elucidating fundamental physical properties and in developing the theory for Hitemperature superconductors (HTSCs), although theory thus far has been unable to account for all verified experimental observations. In establishing the electronic structure of complex cuprate superconducting compounds, the United States is sharing its forefront position with Japan and Germany. The discovery of new superconducting materials around the world has involved as much luck as it has deliberate science-based research. Leadership in this area could shift rapidly when the next important compound is discovered. The United States has enjoyed leadership in the development of magnetic phase diagrams of HTSCs and in the modeling flux pinning and critical phenomena. However, Europe also has shown particular strength in statistical mechanical modeling and in the development of direct imaging techniques for delineating flux line patterns. A relative decline in the fraction of peer-reviewed papers in this field by US investigators is an indicator of the growing competition in basic research world wide.

On the technological front, the United States and Japan appear to be neck-and-neck in most important development areas, especially in bulk superconducting cables, thin-film devices, instrumentation, and power equipment. In technologically important areas, however, it is difficult to assess the degree of US leadership; many advanced developments are being conducted as joint ventures between US companies and their partners in Europe and Japan. Also, many such important cutting-edge developments are closely held.

Early successes in Japan with melt texturing and powder-in-tube processing are now being matched and perhaps exceeded in the United States, but these methods could be replaced by new approaches. For example, a major development in the US is the processing of long-length conductors based on the deposition of highly-textured YBCO on textured metal substrates. Leadership in this area could shift rapidly to the research who successfully replaces plasma laser deposition or evaporation with a low-cost, high-rate process involving metallorganic or chemical vapor deposition. This is an area in which industrial effort dominates in Japan; US industry developments are strongly coupled to work at the national laboratories.

Leadership in the development of rare-earth HTSCs for levitation or energy storage applications is currently shared by the United States and Japan, with Japan in front based on successes with the Nd-123 system. Japan has targeted this area to replace low-temperature superconducting magnets with trapped-field magnets for levitated-train designs.

Electronic application of thin-film superconductors has enjoyed considerable success in the United States because of the strong interest in the use of HTSCs for radio frequency and microwave filters for communications. US leadership in this area has been largely aided by DOD investments in the development of HTSCs for military communications, sensing, and computing.

Research challenges are in the development of phase-pure materials, processing issues, magnetic phase diagrams, electric current transport, and the modeling of transport and critical phenomena. A continuing challenge is the development of a successful comprehensive theory for high-temperature superconductivity.

4.2.8 Polymers

In general, research in polymers is at an exciting stage, and the United States enjoys a strong world position in many areas. The field of polymers is broad and its foundations span from basic chemistry for synthesis to mechanical engineering for processing. Most early research, except some done in academic laboratories in Europe, was done by industry. The field continues to be dominated by chemists and chemical engineers and has only recently been brought under the umbrella of "materials" within the United States. This trend offers a great opportunity for multi-disciplinary research and education.

Polymer research is done in many US universities, both within stand-alone departments and degree programs and by individual investigators who are often found in departments of chemistry or chemical engineering. Academic research in the United States compares well with the rest of the world, despite some powerful research institutions abroad: The Max Planck Institute in Mainz, Germany, for example, enjoys generous long-term funding. Many universities in Japan have strong centers of long standing, for example, at the University of Kyoto. Others are rapidly developing in China, Korea, Taiwan, and Brazil. University-based polymer research in the US scaled to commercial activity is small in comparison with many other university thrust areas.

Leadership of the United States in polymer research is tied to developments in industry and the great economic importance of polymers. Several US companies have had centers of excellence in polymer research, and many products currently on the market were discovered and developed in these laboratories. These laboratories were the envy of the world, and they have been widely emulated. Industrial polymer research is still strong in the United States, but trends of downsizing and a shorter term focus are taking their toll on the rate of innovation; perhaps more so here than abroad.

Among the hot topics is the revolution in the polyolefin industry with the advent of metallocene catalysts that permit unprecedented control of molecular structure and size. Large investments are being made in new manufacturing as a result of this research. The activity is global, but the United States is in a strong leadership position. There is strong research in multicomponent polymer systems around the world, and the United States is a leader. Much of the work is being done with strong university–industry interactions because of the important commercial value of such products. Other forefront topics are biosynthesis, free-radical polymerization, multicomponent systems such as ceramic–organic copolymers, and three-dimensional polymer dendrimers. Major growth areas for polymer applications are separation media, barrier coatings, and packaging; biomedical uses, such as drug delivery and implants; and electronic–photonic applications, such as displays and resists.

4.2.9 Catalysts

Catalysts are central to petroleum refining and chemical processing. They also are used widely in environmental protection technology. In the United States, the value of fuels and chemicals derived from catalysis is nearly 20% of the gross national product (CMR 1997). The annual global market for catalysts is $8 billion (Rothman 1997).

There have been 3 advances in this field of great significance in the past decade: the development of shape-selective catalysts, the development of metallocene catalysts for polymerization, and the application of catalysts for automobile emission control. Shape-selective catalysis in microporous solids is established for many industrial processes in the chemical and petroleum sectors, and the search continues worldwide for new catalysts and applications. The remarkable advances in metallocene catalysts research in the United States now used for the precise control of polymer properties in US industry are spilling over to university-based activities worldwide. The area of environmental catalysis has matured in the past 20 years. New environmental regulations adopted here and in Europe during the past decade have created a need for focused research to support technology development. Catalysis work at industrial laboratories remains strong and significant, although somewhat reduced. The area most affected is basic catalysis research at corporate research centers. For example, there have been significant losses in the petroleum sector; corporate laboratories have been closed and basic catalysis research activities have not been transferred to other divisions of these companies.

Numerous technical societies feature catalysis-related topics among their symposia, and stand-alone societies meet regularly worldwide. The North American Catalysis Society draws about 900 participants at its biannual meeting. About 40% of the papers submitted are from abroad; the European catalysis meetings typically have less than 10% US content. The base for all these meetings is work from university and industrial laboratories.

Several "catalysis centers" have been organized at US universities, although they are not large activities and many have shrunk during the past two decades. The university base for catalyst research has become more diffused as the large university centers have declined in size and as small research groups have been formed at several other universities. Catalyst research, which is generally found in university chemistry and chemical engineering departments, has not benefited as significantly as have other areas of materials science by the migration of researchers from industry to university. Catalysis research is currently more prevalent in chemical engineering departments than it is in chemistry departments in the United States. In some cases, meaningful catalysis research is difficult to conduct in universities because of the laboratory requirements.

The United States is, and will likely continue to be, among the world leaders in industrial practice—particularly in the area of selective alkane oxidation. Although strong and viable, the US industry is small, but it is also small in the rest of the world. Support for catalyst research by companies is not nationally driven; many companies are multinational. Companies locate the catalysts they need wherever they are found in the world. Published reviews of progress in catalytic technology in Japan and Europe provide detailed information of catalytic processes developed in these areas (Roth, 1990).

Catalysis has been an area targeted for growth in most countries outside the United States because of its importance to the industrial sector. European centers of catalysis have grown in recent years. For example, in 1996, the United Kingdom formed an institute to bring its chemical industry closer to academic research (EPSCoR 1996). Similar ventures are occurring in the Netherlands and across Europe. In Germany, funding for catalysis research is associated with the need to sustain development in the chemical industry. In Japan, it has been identified as the motor for green technologies, and increased funding is going to universities. Links between universities and industrial laboratories outside the United States have become stronger than in past years.

Forefront topics in catalysis are selective oxidation, solid acid–base catalysis, novel catalyst characterization techniques, environmental catalysis (for emissions control, waste minimization, reduction of by-products, and evaluation of alternative feedstocks), asymmetric catalysis, and combinatorial catalysis.

5

PROJECTION OF LEADERSHIP DETERMINANTS

This section addresses the questions: "What are the current trends in materials science research in the United States and abroad, and what will the US position be in the near- and long-term future?"

5.1 Overview

Current and future opportunities for materials science and engineering are enormous for several reasons:

- Knowledge of materials and how to tailor their performance economically will be an enabling element for many technologies important to the US economy.
- US research instrumentation and computation facilities are robust.
- There has been a resurgence of interest in processing and synthesis research. This is facilitating the establishment of high-yield, "right-first-time" manufacturing processes, and it has increased the number of pathways for creating new materials.
- The implementation of computational methods is leading to rapid growth in our understanding of complex phenomena and to a reduction in lead time from concept to scientific feasibility.
- The unification of the field and the growth in multidisciplinary collaborations are increasing the productivity and quality of research.
- Graduate education in materials science and engineering is becoming more diverse. It appeals to students with bachelors' degrees in

many science and engineering disciplines, and it provides an array of career opportunities.

No country is in a better position than is the United States to take advantage of these factors. However, several developments could curtail our ability to fully capitalize on our strengths:

- There has been a dramatic change in the Department of Defense (DOD) basic research funding strategy toward areas of strictly military relevance. The generic (dual use) materials research formerly funded by DOD has not shifted to other sources of federal funding. The consequence is a major (~50%) reduction in funds for nonmilitary basic materials research at the universities. This could weaken US leadership in such areas as metals, composites, and ceramics, research areas for which DOD had been a primary federal supporter (Figure 3.9).
- The United States could become less attractive for foreign students and researchers, because of the increasing strength and funding opportunities for materials research and development elsewhere. Having fewer foreign participants could slow research capability; for decades, many prominent materials researchers have come to the United States from abroad.
- A traditional US strength has been the availability of diverse facilities at universities conducting leading-edge materials research. With the decline in federal funding, particularly from DOD, for instrumentation and facilities, complex university laboratories, such as those required for next-generation electronic materials and device structures are having difficulty.
- The elimination of central research laboratories and longer term innovation research by many high-tech companies has made technology transition from universities more difficult. Greater efforts and new pathways are needed to ensure the realization of engineering benefit from new materials concepts developed at universities.
- Shortening the cycle between material invention and engineering application is crucial to continued economic competitiveness in the areas of technology that depend on new materials. However, even though there are considerable opportunities, research funding in this area is limited. Other countries are pursuing this opportunity more aggressively than is the United States: There is a growing partnership between industry, university, and government laboratories to exploit new materials concepts more rapidly, for example, at the National Center for Scientific Research in France.

5.2 Recruitment of Talented Researchers

Talented researchers are needed at all levels in materials research. From the 1950s through the 1980s, US institutions attracted the world's best scientists as principle investigators because of the presence of outstanding researchers with whom these individuals could work, a superior economy, and outstanding research facilities. More recently, there has been considerable global leveling in research capability, and the United States has lost appeal among highly talented scientists in the industrialized world. As shown in Figures 5.1 and 5.2 and Table 5.1, the number of scientists and engineers coming to the United States declined by 26% from 1993 to 1994 (NSF 1997). More current data are needed to determine whether this was a one-time occurrence attributable to immigration from the former USSR before 1994, or a long-term trend.

Although the attraction of talented scientists from less well developed countries will continue, foreign nationals working as materials scientists within the United States are now being heavily recruited by their native countries. Europe, Korea, and Taiwan are enticing scientists working in the United States to return home, and these countries also have begun to attract American researchers. This loss of talent is somewhat offset by the globalization of research and development, permitting US corporations to hire outstanding young scientists at offshore research and development locations. Driven by economic factors and global competitiveness, the trend toward establishing US research facilities abroad is expected to continue. Reciprocal establishment of foreign research facilities within the continental United States has occurred, and continues to be dynamic, but the magnitude of researchers involved may still be overwhelming the system.

There are at least 5 issues that affect the future ability of materials programs to attract high quality graduate students.

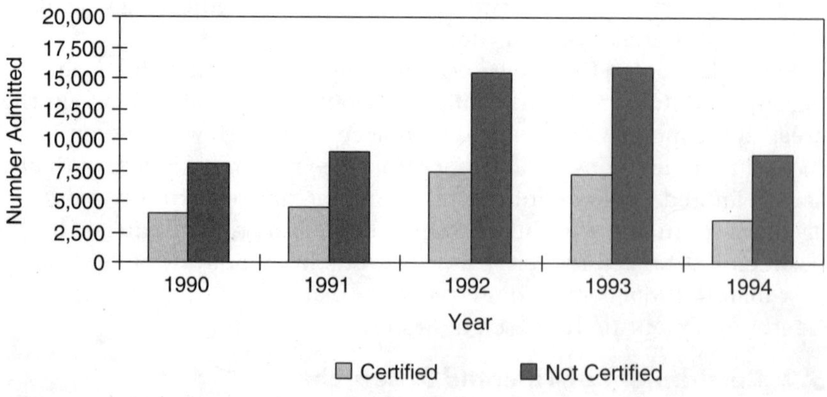

FIGURE 5.1 Scientists and engineers admitted to the US on permanent visas by labor certification, 1990–1994. Source: NSF, 1997.

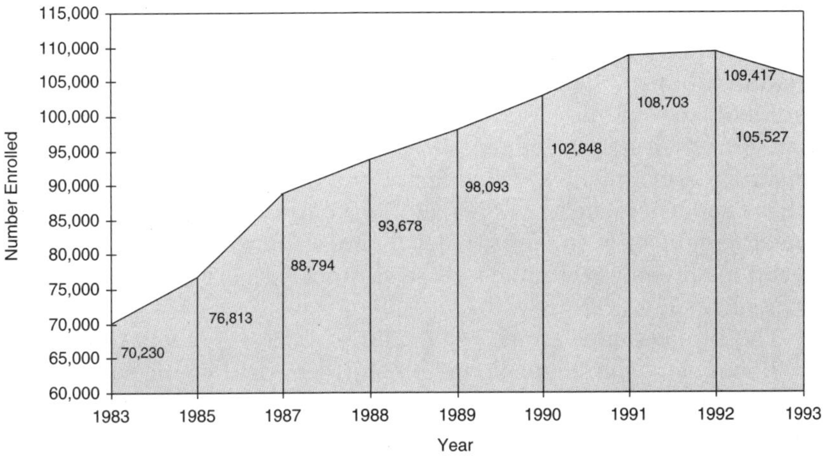

FIGURE 5.2 Foreign citizen graduate enrollment in US science and engineering universities, 1983–1993. Source: NSF, 1996.

First, there is a continuing need to recruit students from other engineering or science fields. For example, science and engineering research in electronic and optical materials has benefited greatly from interdisciplinary work. Recruitment will be facilitated by a broader national recognition and acceptance of materials science and engineering as a distinct academic discipline. This acceptance seems less prevalent here than it is in some other countries.

Second, foreign countries—Korea, Taiwan, China, India, Singapore, Hong Kong—have been the source of graduate students for US programs in many fields, particularly materials science and engineering. Most of these countries have been investing heavily to improve their own

TABLE 5.1 Decline in US Admissions of Immigrant Scientists and Engineers, FY 93–FY 94

Occupation	FY 1993	FY 1994	Percentage change
Engineers	14,497	10,793	-26
Natural scientists	3,901	3,104	-20
Mathematical scientists and computer specialists	4,157	2,781	-33
Social scientists	979	725	-26
TOTAL	23,534	17,403	-26

Source: NSF 1997.

academic programs, especially in the various subfields of materials deemed important for the health of their national economies. The recent economic upheaval in Asian countries is likely to affect the flow of students to the United States—although how or to what extent is not clear. Moreover, not all countries in Asia are at the same stage of industrial development, so the drain of bright young people from some of these countries, such as Korea and Singapore, could soon abate. Thus, the number of students in developing countries who seek graduate education abroad might not decrease as quickly as projected, but the mix could change.

The United States should be concerned, however, about the degree to which it can continue to compete to attract bright young people from abroad. For example, Korea recently began to accept postdoctoral fellows from China. Japan has been encouraging its universities to accept postdoctoral fellows and PhD students from other Asian countries. The same is true for Europe, where there are programs among European Community countries at the postdoctoral and professional levels. One can expect increasing competition for talented students by all developed countries around the world, and the expected demographic decrease in the number of young people in developed countries will make the competition for the brightest even tougher. The greater the degree to which the United States continues to make efforts to make its society active, open, and attractive, the better its chances of attracting a significant fraction of talent pool.

As other countries become more successful in retaining or repatriating their top scientists and engineers, there will be a need to replace them by attracting more US students into PhD programs in materials science and engineering.

Third, the supply of graduate students is directly related to the economy. Few US students pursue graduate education without financial support, so a decline in research funding results in fewer students' pursuing graduate degrees. Moreover, a strong job market for those with bachelor's and master's degrees, particularly for people with an interest in materials, decreases the pool of PhD candidates. These effects are now acting in concert in the United States.

Fourth, materials science and engineering researchers, particularly in industry, must keep constantly up with new developments and research opportunities in the field. Graduate fellowships and educational programs conducted either on-site or "beamed" to industry via satellite from universities are a critical element of leadership.

Fifth, attracting more women and minorities into graduate programs in materials science and engineering is not only essential to achieving greater diversity in academia, government, and industry leadership, but it is also an important means of offsetting our current dependence on foreign sources of PhD talent. The field should be able to benefit from the variety of undergraduate preparations that might attract a greater diversity of talent into graduate materials programs.

5.3 Funding

Whereas US industry and government are shifting funds toward short-term research, many other countries, notably Japan, are increasing long-term and basic research funding. Many US companies have eliminated corporate or central research laboratories to more closely align research and development with immediate business opportunities.

Materials research in universities has been sponsored mainly by the federal government. The mission-oriented agencies (principally DOD and the DOE) have provided the most support for a range of fundamental materials research. In particular, DOD has supported about 60% of *academic* research in the field. The National Science Foundation funding for basic materials research continues to be available (Figures 3.6–3.8) but the recent shift in DOD funding has placed more emphasis on topics of strict military relevance. The consequence will be a curtailment of federal funding for new dual-use materials concepts and small-group research efforts formerly supported by DOD. Some of these new concepts are essential to DOD and to nondefense sectors of the US economy.

A rebalancing of the overall federal research and development funding strategy could be needed to enable materials research in all areas to continue: Otherwise, important developments in key subfields could lag behind in world competition. Many university researchers have had positive experiences with industrial research collaborations.

Although some academic researchers have turned to industry for financial support, in many cases, industry-funded research is of shorter duration and, compared with federal grants, has a specific, short-term focus. Some research projects are conducted under contract terms that capture intellectual properties, protect confidentiality, restrict publication, and require detailed planning and reporting of progress. These conditions rarely attract top graduate talent to the research effort.

The areas in which industrial research collaborations can be most valuable are materials synthesis and processing, where special equipment not generally found in universities is required to achieve process control and to evaluate sequencing protocols and scaling parameters.

5.4 Infrastructure

The quality of the basic research infrastructure and the development of new technology from research strongly influence the long-term health of materials research. The position of the US research enterprise will be determined by the elevation or decline of this infrastructure, which, in this context, is defined broadly to include tangible (facilities) and intangible (supporting policies and services) elements. Several trends for the elements of this infrastructure have been identified:

- The university structure in which the materials science and engineering organization resides strongly influences the fortunes of the discipline. The high quality of academic leadership in materials science and engineering and the excellence of the scientific research enterprise have placed the discipline in a position of strength at most of the top research universities in the United States. The prominence of materials science in nonacademic institutions (industry and government agencies) is also well established here and abroad.
- Maintaining the high scientific and engineering quality of management of materials research organizations within industry and government agencies is critical to the US competitive position. These organizations contribute to the knowledge base through basic research in technologically relevant areas.
- The most advanced information and communication networks available are being used in materials science research. Internet and video conferencing, electronic journals, international distance learning, and distant collaborative research are commonplace. The convergence of voice, data and video systems have made possible the routine use of real-time 3-D imaging and other supercomputing aids for modeling and simulation of materials structures, interfaces, phenomena, and behavior during processing.
- Sophisticated characterization instruments and processing facilities are essential for advancement in materials research. US facilities have instrumentation that is on par with the best in the world. However, rapid advances in design and capabilities of instrumentation can create obsolescence in 5–8 years.
- The integration and overall quality of the characterization services that support US universities and industrial organizations has lost substantial ground to organizations in Japan and Europe. Fortunately, characterization facilities at US national laboratories are among the best. Also, an increasing number of high-quality commercial laboratories are becoming accessible to academic and industrial researchers.
- Small-scale equipment for materials synthesis and processing in most US universities is not keeping pace with similar equipment at some universities abroad. Capabilities in US industry for supplying bulk single crystals and other specialty research materials also have declined. As a result, US researchers are becoming increasingly dependent on foreign sources.
- Forward-looking intellectual property policies, administrative support, and access to patent expertise are improving for US academic researchers in materials science. These policies are generally more flexible and advanced here than they are abroad. The anticipated

continuing liberalization of rules that permit academic researchers to commercialize their inventions is a positive step toward decreasing the time from invention to market. Another positive step is the growing assistance from the universities in finding industrial commercialization partners.

- A supportive public (and thus legislature) is a valued element of the intangible infrastructure. Educating the public about the importance of materials should receive increasing attention. Innovative, attention grabbing methods are needed to convey that everything is made of something, whether a natural or synthetic material. Highlighting materials research contributions to major national initiatives is valuable in sustaining public support for the field.
- Federal laboratories and the national laboratories of DOE are critical in providing unique facilities for research, they have instrumentation no single university could afford to put in place. An important complement is the availability of world-class scientists who engage in long-term fundamental research, provide assistance through research collaborations with the user community, and provide advanced instrumentation design and methods.

Although the US has enjoyed a research and funding environment that allows for the installation and operation of a diverse range of facilities to support leading-edge research in materials, this position is not assured forever. There are two major challenges that must be addressed with a systematic policy.

- Requirements for fabrication and processing facilities will change as research emphases evolve. For example, even though the microelectromagnetic systems (MEMS) field has become active, very few sites are equipped to fabricate MEMS devices, especially in support of the science and engineering research communities. Therefore, most of the materials community is excluded from research opportunities in this field. There are other examples in films, coatings, composites, and integrated sensors. Greater attention is needed to making the best fabrication and processing facilities available to top research teams. In some subfields, this problem has been addressed by collocating fabrication facilities and research teams, such as the nanofabrication facility at Cornell University. In other instances, flexible foundries, such as the Jet Propulsion Laboratory's Metal-Oxide Semiconductor implementation Service, have been made available to top researchers around the country.
- Large central facilities, such as neutron and synchrotron sources, electron microscopy centers, and analytical facilities, many of them at DOE laboratories, must be continuously upgraded and maintained. Funding trends and changing priorities for federal agencies and

NSF raise concerns about whether large-scale facilities can keep pace. In some areas, such as neutron scattering, US facilities have not kept up with foreign competition.

5.5 Cooperative Government–Industrial–Academic Research

Maintaining a competitive advantage in materials science depends on strong collaborations between government, industry, and academia. As industrial research focuses even more on materials technologies with short-term (2–3 year) technology–product impact, execution of longer term (5–10 year) basic and innovative exploratory research at universities and national laboratories will require even closer interactions. Basic research in these areas is a vital aspect of knowledge-based materials science, as verified by the continued university hiring of researchers with industrial experience. Collaborative research is accomplished in several foreign countries by individuals with joint academic–commercial appointments and through publicly supported research institutes linked to universities (similar to many US national laboratories) that serve industry's need for longer term research.

One challenge is also a major opportunity for a government–university–industry initiative: There is a 15-year cycle time in many cases from the scientific feasibility of a new material to its engineering implementation. There is a need for continuity of support and a general recognition of the time it takes to go from observation to hypothesis to experimentation to discovery to implementation. A reduction in this schedule could be realized through modeling and simulation, as applied to fabrication, processing yields, performance, and reliability. There are clearly defined, mutually supportive roles for academia, government, and industry where they can work together. For example, the US semiconductor industry has set up a 15-year road map, and such initiatives are in place in other countries. The DOE advanced-supercomputer initiative is a similar effort to develop new computer methods for the simulation of nuclear weapons.

Industry interest for cooperative programs is strong, but direct industry financing seems impossible to organize. Federal funding by explicit policies does not address this issue, nor is such activity subsumed into ERCs, STCs; and MURIs (see earlier discussion on centers) for which materials research and development is complementary to principal scientific and technology goals. Center programs are needed to put this capability in place in the United States. Without better organized government–industry–university efforts in this area, new materials will be more effectively exploited by other countries.

A novel approach that deserves more widespread use is the establishment of virtual industrial–academic institutes or centers for materials technology development. Some models already exist outside the materials field—in manufacturing, food processing, and biotechnology—

that allow complex, high-risk, long-term basic research in areas with tremendous technological potential to be attacked synergystically. Establishment of new private–public sector partnerships to fund virtual centers would be helpful.

6

LIKELY FUTURE POSITIONS

6.1 Introduction

The likely future position of materials science and engineering in the United States based on the "world congress" assessment, is that our current position among the world leaders is likely to slip in some areas. The reasons vary, but some common elements include the globalization of research and the growth of economies.

The United States can be expected to continue supporting materials science, and new topics involving nanostructure materials and intelligent materials are among several exciting emerging areas of study. In addition, the United States will never want to lose its current strength in aerospace and defense, which are important not only to US security but for the stability of the post-cold-war world. The combination of global threats and economic opportunities can be expected to continue to drive US materials science and technology. Therefore, the panel finds that US leadership position in materials science and engineering should continue.

6.2 Biomaterials

Our strength relative to other countries in basic and applied biomaterials research is likely to erode in the near- and longer term for several reasons. Our lead in basic research is being contested by huge investments in Europe and Japan in biomaterials, both in academia and in industry. The potential market for biomaterials is larger outside the United States than it is within. There is a concern that our lead in applied areas is being jeopardized because of inconsistent and excessively conservative government regulatory policy and because of the litigious climate of our culture in general, which inhibits development and innovation. Although the United States is currently among the world leaders in contemporary diagnostic systems, we are rapidly losing ground to other nations.

6.3 Ceramics

In ceramics used for their thermal, electric, and mechanical characteristics, the United States and Japan share leadership. The Japanese manufacturing advantage is having an effect on engineering and research strengths, and the relative US position is in decline. However, emerging US research strengths in electromechanical systems and coatings would appear to redress the balance on the research side. Japan and Germany continue to be highly competitive in engineering of ceramics.

In areas concerned with functional–electronic ceramics, work on such topics as self-assembly materials and multilayer ferrite processing, where the United States is at the forefront, should be targeted. Areas where we are among the world leaders and where we should maintain our position in the future include three dimensional nanoporous silicates, microwave dielectrics, and electrophoretic preparation of thin films. The United States is not expected to seriously challenge the Japanese leadership position in integrated micromagnetics.

6.4 Composites

Basic research into composites at US universities is coming to a standstill as a result of the Department of Defense decision to strictly curtail university research funding in metal, polymer, and ceramic matrix composites. If this situation long persists, the US could forfeit its leadership role in composites. Academic research is at a nadir, but important new developments in industry are spurring implementation, especially in transportation and construction applications.

6.5 Magnetic Materials

The US is catching up with the leaders in international research on magnetic materials and magnetism. University research in the magnetic materials area has begun to increase. More basic research on magnetic materials and magnetism is needed to increase the prospects for advances by the United States in this area. The vitality of magnetic recording and the phenomenon of colossal magnetoresistance are starting to produce a renaissance in fundamental magnetism research in the United States.

6.6 Metals

In all probability, the US lead will remain, but that is not a certainty. For 5 decades, investments in materials research and development have been driven largely by national security needs. This has yielded a wealth of basic knowledge and new products. Because the United States is the only remaining "superpower", this driving force should remain, but it will diminish in proportion to perceived threats to our national security. This will shift some of the burden for materials development to nondefense industries, such as transportation. The value of new materials to various industries varies widely—and so does support for materials research and development. Another force that will affect the US position is the consolidation and globalization of all industries from aerospace companies to automotive suppliers. For these businesses, the issues of US competitiveness and research and development leadership are much less important, because their playing field is the world and they will seek knowledge wherever it is to be found.

6.7 Electronic and Optical–Photonic Materials

Research in electronics will continue to focus on materials and processes, and it will be conducted globally through international collaborations among industrial organizations. Industry partnerships with academia and government continue to be encouraged by the US Semiconductor Industry Association, which serves as a vehicle for forming other organizations—SEMATECH (the Semiconductor Research Corporation) and MARCO (the Microelectronics Advanced Research Corporation). Such partnerships, along with focused centers at universities, will continue to be vitally important to our leadership in the now-global semiconductor industry.

As semiconductor technology approaches the 100 nm generation, unprecedented new materials and processing technology will be required. The United States and others will make these advances more or less equally, if not as partners. A global forecast for 2012 for technology-driven research directions in this industry for the United States is provided in the National Technology Roadmap for Semiconductors (SIA, 1997). Some of the expected materials and process-related research includes silicon–germanium devices, new gate dielectric and gate electrode materials to replace SiO_2, copper interconnection to replace aluminum, new low-dielectric constant materials for interlayer dielectrics, diffusion barrier materials and processing approaches for copper interconnection, and the replacement of polysilicon gates with high-conductivity gates having low or no depletion. New processing, characterization, and metrology methods will be required to achieve surface smoothness of ±2 Å RMS, gate dielectrics of 10 Å, and unprecedented cleanliness control.

The United States should continue its leadership position in compound semiconductors (GaAs, GaAlAs) and wide-band-gap semiconductors (SiC) for power devices and microwave transmitters. These technologies, which will continue to find expanding applications for cellular telephone and satellite communication systems, are now and should continue to be well funded by industry and the government. Europe should continue to share leadership with the United States in electrical power distribution and motor control applications of power transistors.

In the field of nanotechnology, the United States has traditionally been leading in exploratory nanostructures, including quantum wires and dots. It shares the lead with Europe and Russia in mesoscopic physics.

Although continuing to provide leadership in basic research, the United States has conceded commercial leadership in wide-band-gap photonics to Europe and Japan, and the Japanese currently enjoy a commanding lead in GaN technology and the commercialization of photon-pumped, phosphor-coated ultraviolet emitters for displays. There are many excellent US university capabilities in wide-band-gap photonics, but much of the federal funding for this research is short-term oriented and lacks focus or a coherent long-term strategy to regain leadership in this important field. The Japanese are expected to dominate flat panel display technologies well into the future; US developments in thin-film electroluminescent displays on floppy polymer films could provide an important position in this display market for the United States.

Research support in II-VI (ZnSe) wide-band-gap lasers is being shifted in US universities to nitride research. The Japanese continue to make major strides in improving the longevity and external efficiency of II-VI lasers and LED's. The Japanese lead in short wave length solid state laser communications and in related technologies. The United States leads in longer wave length lasers.

6.8 Superconducting Materials

The current strong position of the United States in superconducting materials is not assured. Many early startup companies that showed promise are not yet profitable. Also, there is still less industrial research here than there is in Japan. Still, some small US companies maintain world leadership in the design, manufacture, and characterization of long-length conductors, although the shift in US corporate research away from longer term basic studies presents a question for the future: "Will the private sector benefit commercially from new concepts and advances occurring at universities and national laboratories?" This field could be one by which the success or failure of government–university–industry partnerships will be judged in the future.

The shift in funding and research priorities among government agencies is affecting federal spending for basic research in favor of advanced developments. The Department of Defense still maintains a program in high-temperature superconductor applications, but total funding for basic research is declining and is increasingly in competition with other funding priorities. Momentum favors relative improvements in the US leadership position in some areas (magnetic properties, flux transport measurements and imaging, thin-film processing, and cable development), but without continued strong federal investment in basic and applied research, that position will change.

On the bright side, dramatic advances in high-temperature superconductors have renewed interest in conventional superconductors for advanced applications, such as SQUIDS, energy storage, instrumentation, and telecommunications. With the improvements to and cost reductions projected for cryocoolers, the market potential for superconductor materials and related technologies should grow. The United States is well poised with strong processing and manufacturing capabilities and a growing talent pool to capture a substantial segment of these markets.

6.9 Polymers

The United States has given less attention and funding than have many other countries to polymer research. Overseas education, research, and infrastructure have had infusions of funding that could change their relative positions, and the United States could lose ground in relative terms if not in absolute terms.

The economic importance of synthetic polymers is evidenced by their wide use in plastics, fibers, adhesives, and paints. These materials are a large fraction of the "chemical" category in which the United States has a strong positive trade balance. Sustaining this balance will require the United States to maintain world leadership in polymer research. Environmental and life cycle responsibility is a driver for polymer research and development in Europe and is becoming more so in the United States.

6.10 Catalysts

The leading position of the United States relative to the rest of the world in the subfield of catalysts is likely to lose ground as a result of the targeted funding aimed at growing capabilities in other countries. This area could stagnate in the United States without stronger, better equipped research centers where researchers can work together with common goals. The *Catalyst Technology Roadmap Report* (Sandia 1997) cites as "the three most important areas of application of catalyst technology in which improvement of catalytic processes would make the most significant progress selective oxidation, alkane activation, and byproduct and waste minimization."

In industrial practice, the United States will remain a world leader for production of chemicals through catalytic reactions in the most energy-efficient, safe, and environmentally compatible way. University-based research will continue to suffer relative to industry laboratories unless better equipment becomes available.

As emerging markets in Asia and Eastern Europe develop and grow, the demand for basic chemicals and polymers (often made catalytically) will grow at double-digit rates. US industry will benefit from this growth because the technology to produce these materials often is not accessible to developing nations, and the expense required to build large-scale manufacturing plants is prohibitive. It is important to continue to invest in research on catalysts and catalysis to promote continued growth of US companies in export markets.

Inventions that use new catalysts or new engineering process concepts that facilitate economic development of small-scale manufacturing plants will allow US companies to invest in new places overseas to serve local markets and speed growth of developing nations. The economies of scale that result from large-volume manufacturing plants support Western economies and allow a cost of manufacture that provides an attractive return on the investment required to build large-volume plants. US companies will continue to seek ways to develop small-scale manufacturing. For example, the concept of "a plant on an IC chip" could some day allow attractive investments by US companies in local bulk chemical and polymer facilities in developing regions of the world.

These points simply argue for continued investment in catalysis research in a way that encourages innovation and allows US industry the flexibility to participate in the growth of emerging markets. Innovation can be encouraged by collaborative research across disciplines and between the public and private sector. Attracting more chemistry students into catalysis research aimed at discovering new science, while simultaneously strengthening links to technology and applications in the engineering area, would benefit the field. A well-funded national institute for catalysis research could serve as the focal point for such collaboration.

7

SUMMARY AND CONCLUSIONS

The report summary and conclusions are provided below. Overall, the analysis was limited by a paucity of field-specific and international data. Nonetheless, the members of the Panel have confidence in the conclusions provided below.

7.1 The United States is among the world's leaders in all subfields, and it is the leader in some.

The United States is currently among world leaders in all of the subfields of materials science and engineering, and currently it enjoys a clear lead in biomaterials. The United States is expected to maintain its lead in metals and electronic–photonic materials because of their large US industrial base. However, the lead in electronic–photonic materials is endangered because of cutbacks in exploratory research. Our earlier preeminence in magnetic materials is now shared with Europe and Japan. This will require particular attention in the future.

Erosion of US leadership is expected in the subfields of composites, catalysts, polymers, and biomaterials because of the high priority being given to these subfields by other countries. Current US weakness in materials synthesis and processing relative to Europe and Japan is especially highlighted in the panel's assessment. In the subfield of catalysts especially, university multidisciplinary research centers with close industry collaborations are needed to conduct cutting-edge research and to reduce the development cycle time for commercialization. Finally, sustaining current federal research support in functional ceramics and superconducting materials is considered important to maintain US leadership in these subfields.

7.2 The flexibility of the enterprise is as much a key indicator of leadership as is the amount of funding.

Funding is important in supporting leadership in materials science and engineering, but a balance among all determinants is required to sustain leadership. Several factors provide opportunities to do this. These include the availability of many options for funding research and entrepreneurial developments through our national innovation processes, a robust infrastructure of research instrumentation and computational facilities, growing opportunities for diversifying the US talent base, and continuing improvements in research quality and productivity through greater unification of the field and growth in multidisciplinary collaborations. Of these opportunities, talent diversity needs much greater effort if it is to be realized.

7.3 The innovation system is a major determinant of US leadership.

The keys to US leadership in the subfields of materials science and engineering have been the entrepreneurial ability of its researchers and the influence of its diverse economy. The rapid exploitation of new developments is facilitated by the extensive networks and collaborations among leading US researchers that extend to all sectors of our economy and throughout the world. The mobility of graduate and postdoctoral entrepreneurs from the academic world to the private sector is stimulated by the availability of venture capital for small start-up companies. Federal programs that encourage research consortia and partnerships in the private sector and that fund precompetitive research at small and medium-size companies provide additional impetus to the development of innovative materials technology.

Flexibility confers agility among US researchers who have been competitive in emerging materials topics—some of which have become "hot" only in recent years. Thorough research can take time, however, and bringing a new material into the marketplace can take more time. The short-term focus of the US innovation system presents the danger of blocking the development of important materials concepts.

7.4 The United States enjoys strength through intellectual and human diversity.

Because materials science and engineering draws from the research infrastructure of most of the physical science and engineering disciplines, it has a high level of intellectual diversity. Intellectual and human diversity are intertwined. As with many science and engineering fields, diversity in educating women and underrepresented minorities in materials science and engineering is now considered inadequate in the United States. However, the disciplines should be especially attractive for diverse representation because of the scope of preparation and technological applications. It is particularly important to attract more domestic students into graduate study even as we continue to recruit the best and brightest from abroad. The ratio of foreign to domestic students at most research universities continues to be high.

7.5 Shifting federal and industry funding priorities, a potential reduction in access to foreign talent, and deteriorating materials research facilities could curtail US ability to capitalize on leadership opportunities.

US leadership in the various subfields of materials science and engineering is not assured for the future. In contrast to opportunities of leadership, there are current developments that could curtail the ability of the US to capitalize on these opportunities. These include shifting materials research and development priorities in the Department of Defense, which have created research gaps in some materials subfields (e.g., ceramics and composite materials, electronic and optical materials), potential decreases in the supply of foreign graduate students, elimination of central research laboratories by major high-tech companies, and lack of attention to research into methods for shortening the implementation cycle for advanced materials.

The US education system—undergraduate and graduate—has achieved excellence that is acknowledged throughout the world, and we continue to attract top talent from other countries, especially those that lack adequate graduate research systems for training research leaders. There is a concern that improvements in graduate education programs in developing countries will not only meet their own needs for building stronger indigenous research, but will attract home the top researchers and students who currently reside in the United States.

One area of special concern is the lack of adequate funding to modernize major research facilities in the United States. Some US facilities are a generation older than are those in other countries, and there are fewer improvements or new facilities being planned for sources of neutrons, synchrotron radiation, and high-energy particle beams (electrons and protons).

Also important to top US researchers is the need to modernize smaller scale research equipment at universities for materials synthesis, processing, and characterization. The concern is that in some subfields such equipment at foreign universities now outclasses what is available at most US universities.

8

REFERENCES

Armor, J. N. *Global Overview of Catalysis: United States of America.* Applied Catalysis A: General 139 (1996) 217-228.

Armor, J. N. *New Catalytic Technology Commercialized in the USA during the 1980s.* Applied Catalysis, 78 (1991) 141-173.

Chauvel, A., Delmon, B. and Holderich, W. F. *New Catalytic Processes Developed in Europe during the 1980s.* Applied Catalysis A: General 115 (1994) 173-217.

CMR (Chemical Market Reporter). April 28, 1997, 251(17): SR20-SR21.

COSEPUP (Committee on Science, Engineering, and Public Policy). *Science, Technology, and the Federal Government: National Goals for a New Era*. Washington D.C.: National Academy Press, 1993.

COSEPUP (Committee on Science, Engineering, and Public Policy). *International Benchmarking of US Mathematics Research*. Washington D.C.: National Academy Press, 1997.

COSEPUP (Committee on Science, Engineering, and Public Policy). *An Assessment of the National Science Foundation's Science and Technology Centers Program*. Washington D.C.: National Academy Press, 1994.

CORE (Committee on Optical Science and Engineering). *Harnessing Light: Optical Science and Engineering for the 21st Century*. Washington D.C.: National Academy Press, 1998.

DOE (Department of Energy). *Neutron Sources for America's Future: Report of the Basic Energy Sciences Advisory Committee Panel on Neutron Sources*, DOE/ER-0576P, 1993.

DOE (Department of Energy). *Report of the Basic Energy Sciences Advisory Committee Panel on DOE Synchotron Radiation Source and Science,* November 1977.

Michael Hart, NSLS Chairman, Synchrotron Radiation News, Vol. 9, No. 5, p. 2.

Misono, M. and Nojiri, N. *Recent Progress in Catalytic Technology in Japan.* Applied Catalysis, 64 (1990) 1-30.

NAE (National Academy of Engineering). *Assessment of the National Science Foundation's Engineering Research Centers Program*. Washington D.C.: National Academy Press, 1989.

Nojiri, N. and Iwamoto, M. *Global Overview of Catalysis: Japan*. Applied Catalysis A:

General 139 (1996) 419-428.
NRC (National Research Council). *Materials Science and Engineering for the 1990's.* Washington D.C.: National Academy Press, 1989.
NRC (National Research Council). *Allocating Federal Funds for Science and Technology.* Washington D.C.: National Academy Press, 1995.
NRC (National Research Council). *Advanced Fibers for High Temperature Ceramic Composites.* Washington D.C.: National Academy Press, 1997.
NRC (National Research Council). *Summary Report 1995: Doctorate Recipients from United States Universities.* Washington D.C.: National Academy Press, 1997.
NSF (National Science Foundation). *Human Resources for Science & Technology: The European Region.* NSF 96-316, p. 21-22. Arlington, VA: National Science Foundation, 1996.
NSF (National Science Foundation). *Science and Engineering Indicators 1996.* NSB 96-21. Washington, D.C.: US Government Printing Office, 1996.
NSF (National Science Foundation). Data Brief, vol. 1997, no. 6. Arlington, VA: National Science Foundation, June 8, 1997.
NSTC (National Science and Technology Council). The Federal Research and Development Program in Materials Science and Technology. 1995.
OSTP (Office of Science and Technology Policy). Advanced Materials and Processing: Fiscal Year 1993 Program.
OSTP (Office of Science and Technology Policy). Advanced Materials and Processing: Fiscal Year 1995 Program.
Roth, J. F. *Industrial Catalysis: Poised for a New Generation of Major Innovations.* Applied Catalysis 64, 1-30 (1990).
Rothman, David. Chemical Week. October 15, 1997, p.41.
Sandia (Sandia National Laboratory). *Catalyst Technology Roadmap Report.* SAND97-1424*UC-1404, June 1997.
SIA (Semiconductor Industry Association). *The National Technology Roadmap for Semiconductors: Technology Needs, 1997.*
UK (United Kingdom Office of Science and Technology). *The Quality of the UK Science Base.* London, UK: Department of Trade and Industry, March 1997.

APPENDIX A

PANEL AND STAFF BIOGRAPHICAL INFORMATION

Panel

Arden L. Bement, Jr., is the Basil S. Turner Distinguished Professor of Engineering and director of the Midwest Superconductivity Consortium at Purdue University. Before this appointment in December 1992, he was vice president for science and technology at TRW, Inc. He joined TRW in 1980 as vice president for technical resources. Dr. Bement began his professional career in 1954 as a research metallurgist and reactor project engineer with the General Electric Company at the Hanford Atomic Products Operation, Richland, Washington. In 1965 he joined Battelle Memorial Institute as manager of the Metallurgy Research Department. Three years later, he became manager of the Fuels and Materials Department. In 1970, Dr. Bement joined the faculty of the Massachusetts Institute of Technology as professor of nuclear materials, and in 1976 became director of the Materials Science Office of the Defense Advanced Projects Agency. In 1979, he was appointed deputy undersecretary of defense for research and engineering. In 1990 the US Senate confirmed Dr. Bement as a member of the National Science Board for a term that expired in 1994. Dr. Bement is a member of the National Academy of Engineering, is a recipient of the Distinguished Civilian Service Medal of the Department of Defense, and holds an honorary doctorate of engineering from Cleveland State University. He received an EMet from the Colorado School of Mines, an MS from the University of Idaho, and a PhD from the University of Michigan. Dr. Bement has served on many advisory committees and boards for government agencies and nonprofit organizations.

Peter R. Bridenbaugh is retired executive vice president of the Aluminum Company of America (Alcoa). He also served as Alcoa's chief technical officer and vice president for research and development. Dr. Bridenbaugh holds a BS (1962) and an MS (1966) from Lehigh University, and a PhD (1968) from the Massachusetts Institute of Technology. He is a member of the American Society of Metals, the American Institute of Mining and Metallurgical Engineers, the American Society for Engineering Education, the Industrial Research Institute, and Sigma Xi.

Leroy L. Chang is dean of science and professor of physics at the Hong Kong University of Science and Technology. He was with IBM's Thomas J. Watson Research Center from 1963 to 1968 and again from 1969 to 1992. From 1985 to 1992, he was manager of quantum structures. In 1968–1969 he was an associate professor of electrical engineering at the Massachusetts Institute of Technology. He received the International Prize for New Materials (1985) from the American Physical Society, the David Sarnoff Award (1990) from the Institute of Electrical and Electronic Engineers, and the Stuart Ballantine Medal (1993) from the Franklin Institute. He is a member of the National Academy of Sciences and the National Academy of Engineering and also a member of the Chinese Academy of Sciences.

Daniel S. Chemla has been director of the Materials Science Division at the Lawrence Berkeley National Laboratory and professor in the Department of Physics at the University of California, Berkeley, since 1991. He was with AT&T Bell Laboratories as head of the Quantum Physics and Electronics Research Department from 1983 to 1990; from 1981 to 1983 he was a member of the technical staff, both in the Electronic Research Laboratory. Dr. Chemla has held the positions of department head and group leader with the Centre National d'Etudes des Telecommunications in Bageaux, France (1974–1981). He received a degree as Ingenieur Civil des Telecommunications (1965) from the Ecole Superieure Nationale des Telecommunications in Paris. He holds the Diplome d'Etudes Approfondies, Physique des Solides (1967), and a PhD (1972) from the University of Paris. He is co-recipient of the 1988 R. W. Wood prize from the Optical Society of America. He received the 1995 Quantum electronics Award of the Institute of Electrical and Electronics Engineers/Laser and Electro-Optical Society, and the 1995 Humboldt Research Award. Dr. Chemla is a fellow of the Physical Society of America, the IEEE-Laser and Electro-Optical Society, and the Optical Society of America. He is a member of the National Academy of Sciences.

Uma Chowdhry is business planning and technology director for DuPont's Specialty Chemicals Businesses. Her career since receiving a

PhD in materials science and engineering from the Massachusetts Institute of Technology in 1976 has been with DuPont. She has served in various technology management positions. She managed groups working on ceramic materials for electronic applications, superconductors, and catalysts for various heterogeneously catalyzed chemical processes. In 1995, she was appointed business director for a $400 million chemical intermediates business that commercialized two new technologies involving large-scale catalytic processes. Dr. Chowdhry's technical background is in ceramics processing and in heterogeneous catalysis; her current work is in management of technology for technology-based businesses. In 1996, she was elected membership in the National Academy of Engineering.

Anthony G. Evans is Gordon McKay Professor of Materials Engineering at Harvard University's Division of Applied Sciences (1994–present). Concurrently (1985–present), he is Alcoa Professor and codirector of the High Performance Composites Center in the Materials Department at the University of California, Santa Barbara. Previous to this he was Alcoa Professor and chair of the Materials Department. From 1978 to 1985 he was a professor in the Department of Materials Science and Mineral Engineering at the University of California, Berkeley. Dr. Evans holds a BSc (1964) and PhD (1967), both in metallurgy, from Imperial College in London. He is chair of the Defense Sciences Research Council, a vice president of the American Chemical Society, and a member of the National Materials Advisory Board. He is the recipient of many honors and awards and is a member of the National Academy of Engineering.

Paul Hagenmuller is professor emeritus at the University of Bordeaux (1994) and honorary director of the CNRS Solid State Chemistry Laboratory (1986), which he created in the 1960s. He is editor of many scientific journals and author or co-author of several hundred publications. He is a recipient of two von Humboldt prizes (1997 and 1990), the Gauss-Weber Medal from the University of Gottigen (1997), the Henri Moissan International Prize (1997), and a host of other honors and awards. Dr. Hagenmuller is a member or honorary member of the New York Academy of Sciences; European Academy of Arts, Sciences, and Humanities; the European Academy; the International Academy of Ceramics; and the Materials Research Society of India, among others. He also holds several honorary professorships and degrees. Among his many military decorations is the Croix de Guerre with Palms (1949).

James W. Mitchell is director of materials, reliability and ecology research at Bell Laboratories, Lucent Technologies (1995–present). He has been with AT&T Bell Laboratories since 1970 as a member of the technical staff (1970–1972), supervisor of the Inorganic Analysis Group (1972–1975), head of the Analytical Chemistry Research (1975–1994), research fellow (1985), and head of the Process and Chemical Engineering Research Department (1994–1995). Dr. Mitchell received a BS in chemistry (1965) from the Agricultural & Technology State University of North Carolina in Greensboro and a PhD in analytical chemistry (1970) from Iowa State University in Ames. He received the Percy L. Julian Industrial Research Award (1981) from the National Organization for the Professional Advancement of Black Chemists and Chemical Engineers; the US Black Engineer of the Year Award (1993) from the Council of Engineering Deans of Historically Black Colleges and Universities; the Iowa State University, George Washington Carver Visiting Professorship Award (1994); and the AT&T Bell Laboratories Research Fellow Award (1985), among others. Dr. Mitchell is a fellow of the American Institute of Chemists, a member of the New York Academy of Sciences, the American Chemical Society, the National Organization for the Professional Advancement of Black Chemists and Chemical Engineers, and of the National Academy of Engineering.

Donald R. Paul holds the Melvin H. Gertz Regents Chair in Chemical Engineering and is director of the Texas Materials Institute at the University of Texas at Austin. His research involves structure–property relationships and processing of polymers; his current work deals with polymer blends and membranes. Dr. Paul received a BS in chemical engineering from North Carolina State University and an MS and a PhD in chemical engineering from the University of Wisconsin. He is editor of *Industrial & Engineering Chemistry Research* (published by the American Chemical Society) and serves on the editorial boards of the *Journal of Applied Polymer Science*, *Polymer Engineering and Science*, *Polymer*, the *Journal of Membrane Science*, the *Journal of Polymer Science–Polymer Physics*, and *Polymer Contents*. He has received numerous awards for teaching and research, including in 1988 his election to the National Academy of Engineering.

Buddy D. Ratner is professor of bioengineering and chemical engineering at the University of Washington. He received a PhD (1972) in polymer chemistry from the Polytechnic Institute of Brooklyn. From 1985 to 1996 he directed the National Institutes of Health-funded National ESCA and Surface Analysis Center for Biomedical Problems. In 1996, he assumed the directorship of the University of Washington's engineered biomaterials research center, which is funded by NSF. He is the coeditor of the journal *Plasmas and Polymers*, a past president of the Society for Biomaterials, and author of more than 250 scholarly works. Dr. Ratner is a fellow of the American Institute of Medical and Biological Engineering, the American Vacuum Society, and the Society for Biomaterials. His research interests include biomaterials, polymers, biocompatibility, surface analysis of organic materials, self assembly, and rf–plasma thin-film deposition.

Kathleen C. Taylor is head of the Physics and Physical Chemistry Department of the General Motors Global Research and Development Center. She is responsible for management of research and development in materials science with primary responsibility for 85 PhD, MS, and BS engineers and scientists involved in research programs in exhaust emission control, catalysis, surface chemistry, air pollution control, advanced batteries, fuel cells, corrosion, protective and wear-resistant coatings, light metals, magnetic and optical materials, and chemical and magnetic field sensors. Dr. Taylor was elected to the National Academy of Engineering in 1995, and she is a member of the American Chemical Society, the Materials Research Society, and a fellow of the American Association for the Advancement of Science and the Society of Automotive Engineers International.

Robert M. White is university professor and head of the Department of Electrical and Computer Engineering at Carnegie-Mellon University. Before joining CMU in 1993, he served during the Bush Administration as the first undersecretary of commerce for technology. Before going to Washington, he spent 6 years with Control Data Corporation, first as vice president of research for CDC's Data Storage Products Group and then as the corporation's chief technical officer and as a member of the CDC Management Board. Dr. White's early career was spent in teaching and research. He was an assistant professor of physics at Stanford University and was principal scientist at Xerox's Palo Alto Research Center for 13 years. He is the author of 4 books and more than 100 technical publications on condensed-matter physics and magnetic recording. White received a BS (1960) in physics from Massachusetts Institute of Technology and his PhD (1964), also in physics, from Stanford. He is a member of the National Academy of Engineering and a fellow of the American Physical Society, the Institute of Electrical and Electronics Engineers, and the American Association for the Advancement of Science. In 1980, he received the Alexander von Humboldt Prize from Germany, and in 1993 the IEEE Award for Contributions to Public Service. He is a director of SGS-Thomson Microelectronics, Zilog, and Ontrack Data International.

Masaharu Yamaguchi is professor in the Department of Materials Science and Engineering at Kyoto University, Japan (1987–present). From 1973 to 1987 he was with Osaka University as an associate professor in the Department of Materials Science and Engineering. From 1965 to 1973 he was a research associate in the Department of Metallurgy. Dr. Yamaguchi received his BSc (1963), MSc (1965), and PhD (1969) from Osaka University. He also is Editor of *Intermetallics* for Elsevier Science Publishers. Dr. Yamaguchi received the Meritorious Activity Medal (1983) and the Tanigawa-Harris Medal (1995), both from the Japan Institute of Metals.

Staff

Deborah D. Stine is the study director and associate director of the Committee on Science, Engineering, and Public Policy (COSEPUP). She has worked on various projects throughout the National Academy of Sciences complex since 1989. She received a National Research Council group award for her first study for COSEPUP on policy implications of greenhouse warming, and a Commission on Life Sciences staff citation for her work in risk assessment and management. Other studies have addressed graduate education, responsible conduct of research, careers in science and engineering, environmental remediation, the national biological survey, and corporate environmental stewardship. Dr. Stine holds a bachelor's degree in mechanical and environmental engineering from the University of California, Irvine; a master's degree in business administration; and a PhD in public administration, specializing in policy analysis, from the American University. Before coming to the academy, she was a mathematician for the US Air Force, an air-pollution engineer for the state of Texas, and an air-issues manager for the Chemical Manufacturers Association.

Lawrence E. McCray is executive director of the Committee on Science, Engineering, and Public Policy (COSEPUP). He held positions in the Environmental Protection Agency, the US Regulatory Council, and the Office of Management and Budget before coming to the National Academy of Sciences in 1981. He has directed academy studies in carcinogenic risk assessment, export controls, nuclear winter, and federal science budgeting. A Fulbright scholar in 1968, Dr. McCray received the Schattschneider Award in 1972 from the American Political Science Association for the best dissertation in American government and politics. In 1987, he received the National Research Council staff award. Dr. McCray joined COSEPUP in 1988 as executive director and since 1994 has served concurrently as the director of the National Research Council Policy Division.

Patrick P. Sevcik is a research associate with the Committee on Science, Engineering, and Public Policy (COSEPUP). He works on a variety of projects for COSEPUP, the Policy Division (PD), and the PD Office of Special Projects, assisting Deborah Stine and Lawrence McCray. Before coming to the National Research Council in 1993, he was an assistant program officer with the International Republican

Institute from 1990 to 1993, working on democracy development, primarily in central and eastern Europe. Mr. Sevcik has held positions at the White House in the Office of Political Affairs (1989–1990) and on Capitol Hill (1987–1988) in the office of Rep. John DioGuardi (R-NY). During that time, he also held concurrent positions in several Slovak–American organizations. He holds a BA in international affairs, with an emphasis on Soviet and Eastern European studies, from the George Washington University. He has also studied Russian language and culture at the Leningrad Polytechnic Institute in Leningrad. Mr. Sevcik will begin an MS program in health sciences management and policy at New York Medical College in June 1998.

APPENDIX B

BENCHMARKING RESULTS TABLES

Relative Position of Subfield: Biomaterials

Sub-Subfield	Current Position					Likely Future Position					Comments
	1 Fore-front	2	3 Among world leaders	4	5 Behind world leaders	1 Gain-ing/Ex-tending	2	3 Main-taining	4	5 Losing	
Tissue engineering	●					●					Clear US leadership; tremendous worldwide interest.
Molecular architecture			●					●			Strong US competition from Germany and Japan.
Protein analogs	●					●					US dominates, driven by a basic-science approach.
Biomimetics			●					●			Strong players in North America, UK, Japan.
Contemporary diagnostic systems			●						●		Large European Community investments in biosensors research could lower US ranking.
Advanced controlled-release systems		●					●				US leads; extremely high worldwide interest could change this.
Bone biomaterials			●					●			Important developments in Europe and Japan.

Relative Position of Subfield: Ceramics

Sub-Subfield	Current Position					Likely Future Position					Comments
	1 Fore-front	2	3 Among world leaders	4	5 Behind world leaders	1 Gain-ing/Ex-tending	2	3 Main-taining	4	5 Losing	
Sol-gel-derived materials	●							●			Area advanced by US for production of monolithic glass.
Self-assembled materials	●					●					US leads in fundamental advances, technologic innovation.
Integrated micromagnetics					●	●					Japan leading in power systems on-a-chip applications; US and others ahead in development of new materials.
Multilayer ferrite processing	●					●					Being advanced primarily by US industry.
3D Nanoporous silicates			●					●			New synthesis approach deserves greater scrutiny
Microwave dielectrics			●					●			Worldwide attention focused on producing low-, high-dielectric-constant materials.
Electrophoretic thin-film preparation			●					●			Area ripe for basic, applied materials research.

Relative Position of Subfield: Ceramics (continued)

Sub-Subfield	Current Position					Likely Future Position					Comments
	1 Fore-front	2	3 Among world leaders	4	5 Behind world leaders	1 Gain-ing/Ex-tending	2	3 Main-taining	4	5 Losing	
MEMS Heat engines	●						●				MEMS heat engines made from SiC are a new US discovery. An exciting technology, and US enjoys a major lead. Investments in MEMS fabrication facilities, foundries needed to exploit opportunities.
Single-crystal high-authority ferroelectrics			●						●		Single-crystal-oxide ferroelectrics discovered in US provide unprecedented large strains in concert with large forces (high authority). Exploitation by DOD has begun. Broad-based effort needed to establish commercial technology that benefits from discovery.
AlN–Diamond heat dissipation for power electronics			●						●		High-thermal-conductivity dielectrics, especially AlN, SiC, diamond are crucial for power electronics. Production capacity dominated by Japan; an unstable situation for the US.

Relative Position of Subfield: Ceramics (continued)

Sub-Subfield	Current Position					Likely Future Position					Comments
	1 Fore-front	2	3 Among world leaders	4	5 Behind world leaders	1 Gain-ing/Ex-tending	2	3 Main-taining	4	5 Losing	
Films, coatings (thermal barrier coatings, diamondlike carbon, hydroxyapetite)			●						●		Ceramic films and coatings are increasingly important for thermal protection, wear resistance, corrosion protection. US leads; major efforts in the European Union, Japan.

Relative Position of Subfield: Composites

Sub-Subfield	Current Position					Likely Future Position					Comments
	1 Fore-front	2	3 Among world leaders	4	5 Behind world leaders	1 Gain-ing/Ex-tending	2	3 Main-taining	4	5 Losing	
Polymer matrix composites		●							●		Implementation slow, because of cost. Cost reduction efforts continue, worldwide. Industry activity: US, Japan, France equally engaged.
(a) Large integrated structures		●						●			Use of lower temperature curing matrices and electron beams allows manufacture of large integrated structures needed to reduce cost. Little basic research in US in support; France more progressive.
(b) Ambient-temperature curing (electron beams)			●						●		Enabling for integrated structures: France active.
(c) Design, testing protocols			●						●		New test methods and design practices needed to support cost reduction strategies. Minimal US activity other than NASA–ARL.

Relative Position of Subfield: Composites (continued)

Sub-Subfield	Current Position 1 Fore-front	2	3 Among world leaders	4	5 Behind world leaders	Likely Future Position 1 Gain-ing/Ex-tending	2	3 Main-taining	4	5 Losing	Comments
Ceramic matrix composites		●						●	●		Implementation in energy, aerospace sectors imminent. Research funding at US universities has essentially ceased. France and Japan have major initiatives.
(a) Oxide composites		●						●			Recent emphasis. Most long-life applications required oxides; technology immature. US has slim leadership position.
(b) Nonoxide composites, fibers			●						●		Japan and Germany more proactive. Stress oxidation remains a problem for many applications. Little academic activity on this topic. Japan has new program.
Metal matrix composites		●						●			New applications for particle-reinforce Al alloys led resurgence of US interest in these composites. Minimal research backup.

Relative Position of Subfield: Composites (continued)

Sub-Subfield	Current Position					Likely Future Position					Comments
	1 Fore-front	2	3 Among world leaders	4	5 Behind world leaders	1 Gain-ing/Ex-tending	2	3 Main-taining	4	5 Losing	
(a) Particle-reinforced alloys		●							●		Some research on Al_2O_3–Al materials; research on SiC–Ti has ceased. US remains ahead.
(b) Continuos fiber		●								●	

Relative Position of Subfield: Magnetic Materials

Sub-Subfield	Current Position					Likely Future Position					Comments
	1 Fore-front	2	3 Among world leaders	4	5 Behind world leaders	1 Gain-ing/Ex-tending	2	3 Main-taining	4	5 Losing	
Thin-film micromagnetics of			●					●			CMU, NRL. Germany
Interlayer magnetic coupling	●							●			NIST, IBM
Giant magnetoresistance (spin valves)	●							●			IBM
Spin-dependent tunneling			●			●					MIT, CMU. Japan
Magnetic nanostructures	●							●			Stanford, UCSD
Colossal magnetoresistance		●						●			Univ. of Maryland, many others

Relative Position of Subfield: Metals

Sub-Subfield	Current Position					Likely Future Position					Comments
	1 Fore-front	2	3 Among world leaders	4	5 Behind world leaders	1 Gain-ing/Ex-tending	2	3 Main-taining	4	5 Losing	
High-temperature structural intermetallics		●					●				US among leaders in basic experimental work; at forefront in studies for real structural applications.
Amorphous (bulk), quasicrystalline, nanostructured materials (high-strength materials)			●					●			Bulk-glass-forming alloys were discovered in the US. Intensive study is going on in Japan.
Theory, modeling of atomic bonding, crystal structure, interfaces, phase diagrams, phase transformations, properties	●					●					Ab-initio calculations and non-ab-initio modeling excellent in US. Leadership at US national laboratories and universities.
GMR, related materials	●					●					Excellent studies for applications.
Hydrogen-absorbing materials applications for batteries, hydrogen storage				●					●		Recent intensive studies in Germany, Japan.
Advanced processing of materials to net shape (metallic alloys)	●					●					Excellent work in US superalloys industry (jet engine disks).

Relative Position of Subfield: Metals (continued)

Sub-Subfield	Current Position					Likely Future Position					Comments
	1 Fore-front	2	3 Among world leaders	4	5 Behind world leaders	1 Gain-ing/Ex-tending	2	3 Main-taining	4	5 Losing	
Quantitative understanding and models of plastic deformation (polycrystalline materials)			●					●			Good work in Europe, US in national laboratories, universities, industry.
Quantitative understanding of structure evolution, plastic deformation of polycrystalline metallic alloys		●					●				Strong US and European capabili-ties and programs.
Integration of models of structure evolution, plastic deformation, composition, processing (concur-rent product– process design)		●					●				Good, but generally under–funded, programs in Europe, US. Strong capabilities at national laboratories, universities, and some companies in US, Europe.
Integration of dimensional scales from atomic clusters to test coupons to final products			●				●				No clear leader in this relatively new area.
Net shape, novel processing of metallic alloys		●				●					Will continue to be a major interest of global industries; no clear leader.

Relative Position of Subfield: Metals (continued)

Sub-Subfield	Current Position					Likely Future Position					Comments
	1 Fore-front	2	3 Among world leaders	4	5 Behind world leaders	1 Gain-ing/Ex-tending	2	3 Main-taining	4	5 Losing	
Next generation of high-temperature alloys	●							●			Effects of recent massive changes in global aerospace, defense industries not yet known
Surface treatments to enhance structural performance			●					●			Coatings, etc., widely used; quantitative knowledge of effect on structural performance is weak.

Relative Position of Subfield: Electronic and Optical-Photonic Materials

Sub-Subfield	Current Position					Likely Future Position					Comments
	1 Fore-front	2	3 Among world leaders	4	5 Behind world leaders	1 Gain-ing/Ex-tending	2	3 Main-taining	4	5 Losing	
Deep UV, electron lithography		●					●				US industry leads; rest of the world nearly equal. Key to further miniaturization in innovation a strong US area.
Systems-on-a-chip	●						●				Simulation, modeling extremely critical. US occupies preeminent position.
Copper metalization		●						●			Processing R&D vigorous world-wide.
Submicrometer plasma processing	●					●					US, Japanese industries collaborate.
Holographic storage materials		●					●				US industry, academia lead the world
Organic transistors			●			●					European industry, universities strong; US leads in materials, processing.
Photonic band-gap materials		●				●					US universities, industry lead.

Relative Position of Subfield: Electronic and Optical-Photonic Materials (continued)

Sub-Subfield	Current Position 1 Fore-front	2	3 Among world leaders	4	5 Behind world leaders	Likely Future Position 1 Gain-ing/Ex-tending	2	3 Main-taining	4	5 Losing	Comments
Organic lasers, LEDs			●			●					US, European industry nearly equal; Japan expanding involvement.
Blue-green lasers (gallium nitride materials)					●			●			Japanese industry lead; US industry competitive.
Semiconductor processing			●					●			Comparable in industrialized countries; US, Japan lead.
Interconnects			●					●			Activities mainly in industry. This area needs, and soon could have, important innovations.
Magnetic Storage		●						●			US, Japan share leadership in GMR.
Widegap Lasers and Display	●							●			Japan started in widegap display, led in gallium nitride; other countries, including US, gaining. Japan is the clear leader in liquid crystal display.

Relative Position of Subfield: Electronic and Optical-Photonic Materials (continued)

Sub-Subfield	Current Position					Likely Future Position					Comments
	1 Fore-front	2	3 Among world leaders	4	5 Behind world leaders	1 Gain-ing/Ex-tending	2	3 Main-taining	4	5 Losing	
Nanomaterials	●							●			Frontier with a promising future. US started, maintained lead; Europe, Japan investing heavily.
Semiconductor equipment		●					●				US has advanced recently. Sematech contributed to success.
Wireless				●			●				Strong capabilities in Europe.
Fibers	●							●			US does well in advancing research.

Relative Position of Subfield: Superconducting Materials

Sub-Subfield	Current Position: 1 Fore-front	2	3 Among world leaders	4	5 Behind world leaders	Likely Future Position: 1 Gain-ing/Ex-tending	2	3 Main-taining	4	5 Losing	Comments
High-temperature superconductors (general)		●						●			Leadership between US and Japan
High-temperature superconductor snythesis			●					●			Leadership distributed globally. Could change with next compound discovered.
Processing of highly textured, dense bulk forms for wire, energy storage			●				●				Strong programs at ANL, LANL, and ORNL, US and Japan co-leaders. Japan is especially strong in energy storage.
Magnetic phase diagrams, properties	●						●				Strong capabilities at US universities, national laboratories.
Statistical mechanical modeling of transport and critical phenomenon		●					●				US strong but not dominant; strong European capabilities.
Experimental measurement of flux transport mechanisms		●					●				Strong US university, national laboratory capabilities.
Modeling of optical, electronic properties			●					●			US leads fundamental research.

Relative Position of Subfield: Superconducting Materials (continued)

Sub-Subfield	Current Position					Likely Future Position					Comments
	1 Fore-front	2	3 Among world leaders	4	5 Behind world leaders	1 Gain-ing/Ex-tending	2	3 Main-taining	4	5 Losing	
Physical properties (other than magnetic)		●						●			Strong leadership at US universities.
Development of fluxoid imaging technologies		●					●				Strong capabilities at US universities, industry, and national laboratories. Leading capabilities in Europe.
Thin-film deposition processes		●							●		US leads; Japan could overtake.
Epitaxial, patterning techniques	●						●				US leads in surface, interface science.

Relative Position of Subfield: Polymers

Sub-Subfield	Current Position					Likely Future Position					Comments
	1 Fore-front	2	3 Among world leaders	4	5 Behind world leaders	1 Gain-ing/Ex-tending	2	3 Main-taining	4	5 Losing	
1. Controlled polymerization (a) Metallocene polymerization of olefins (b) Living free radical polymerization (c) Atom transfer radical polymerization (d) Dendrimer polymerization (e) Biologic synthesis (f) Supercritical CO_2 as a polymerization medium		●						●			US leads in most areas; other countries have important programs, especially in (a) and (b)
2. Multicomponent systems (a) Blends or alloys (b) Block, graft copolymers (c) Nanocomposites (d) Macrocomposites (e) Thin-film laminates (f) Interfaces		●						●			US has strong position; many other countries investing heavily.

Relative Position of Subfield: Polymers (continued)

Sub-Subfield	Current Position					Likely Future Position					Comments
	1 Fore-front	2	3 Among world leaders	4	5 Behind world leaders	1 Gain-ing/Ex-tending	2	3 Main-taining	4	5 Losing	
3. Biomedical polymers (a) Implants (b) Drug delivery	●						●				US is preeminent.
4. Electronic–Photonic (a) Conducting polymers (b) Polymers for display devices (c) Resist materials (d) Electroluminescent		●					●				
5. Separation media (a) Membranes (b) Molecular recognition (c) Barrier materials (d) Modified atmosphere packaging (e) Coatings		●					●				US position strong; Europe, Asia have strong efforts in membranes.
6. Theory, modeling (a) Molecular simulation (b) Monte Carlo techniques (c) Conformation (d) Scaling theory		●						●			US is very strong; strong efforts also in Europe.

Relative Position of Subfield: Polymers (continued)

Sub-Subfield	Current Position					Likely Future Position					Comments
	1 Fore-front	2	3 Among world leaders	4	5 Behind world leaders	1 Gain-ing/Ex-tending	2	3 Main-taining	4	5 Losing	
7. Processing (a) Rheology (b) Flow instability (c) Computer modeling (d) New processes			●					●			Very strong efforts in Germany

Relative Position of Subfield: Catalysts

Sub-Subfield	Current Position: 1 Fore-front	2	3 Among world leaders	4	5 Behind world leaders	Likely Future Position: 1 Gaining/Extending	2	3 Maintaining	4	5 Losing	Comments
Catalysis			●						●		Shape-selective catalysis, metallocene catalysis for polymerization, and application of catalysts for emissions control (automobile) economically critical in US.
Selective oxidation				●					●		Selective oxidation is a growing area; applications from small to heavy chemical synthesis (30–40 million tons annually). Industry leaders in US and Europe.
Solid acid–base catalysis			●							●	Industrial activity highly competitive, secretive, largely focused in US.
Environmental catalysis			●							●	Environmental progress requires highly sophisticated industrial work. Advances made concert with applications. Strong capabilities in US, Europe, Japan.

Relative Position of Subfield: Catalysts (continued)

Sub-Subfield	Current Position					Likely Future Position					Comments
	1 Fore-front	2	3 Among world leaders	4	5 Behind world leaders	1 Gain-ing/Ex-tending	2	3 Main-taining	4	5 Losing	
Catalyst characterization			●					●			This area has benefited from advances in atomic resolution microscopy, necessarily equipment dependent. Utility of work depends on strong links to applications.
Combinatorial catalysis	●							●			Still in its infancy; US is strong.
Asymmetric catalysis				●					●		Highly specialized field of great importance limited-quantity manufacturing of products (significantly below the commodity level)— pharmaceuticals, agricultural chemicals. Industry leaders in US, Europe, Japan.

APPENDIX C

HOT TOPICS LIST

Biomaterials

- Tissue engineering
- Molecular architecture
- Protein analogs
- Biomimetics
- Contemporary diagnostic systems
- Advanced controlled-release systems
- Bone materials

Ceramics

- Sol-gel-derived materials
- Self-assembled materials
- Integrated micromagnetics
- Multilayer ferrite processing
- Three-dimensional nanoporous silicates
- Microwave dielectrics
- Electrophoretic preparation of thin films
- MEMS heat engines
- Single-crystal high-authority ferroelectrics
- AlN/Diamond heat dissipation for power electronics
- Films and coatings (thermal barrier coatings, diamondlike carbon, hydroxyapetite)
- Carbon Nanotubes

Composites

- Polymer matrix composites
- Large integrated structures
- Ambient temperature curing (electron beams)

- Design and testing protocols
- Ceramic matrix composites
- Oxide composites
- Nonoxide composites and fibers
- Metal matrix composites
- Particle-reinforced alloys
- Continuous fiber

Magnetic Materials

- Micromagnetics of thin films
- Interlayer magnetic coupling
- Giant magnetoresistance (spin valves)
- Spin-dependent tunneling
- Magnetic nanostructures
- Colossal magnetoresistance

Metals

- High-temperature structural intermetallics
- Amorphous (bulk), quasicrystalline, and nanostructured materials (high-strength materials)
- Theory and modeling of atomic bonding, crystal structure interfaces, phase diagrams, phase transformations, and properties
- Giant Magnetoresistance and related materials
- Hydrogen-absorbing materials applications for batteries and hydrogen storage
- Advanced processing of materials to net shape (metallic alloys)
- Quantitative understanding and modeling of plastic deformation (polycrystalline materials)
- Quantitative understanding of structure evolution and plastic deformation of polycrystalline metallic alloys
- Integrated of models of structure evolution, plastic deformation, composition, and processing (concurrent product–process design)
- Integrated of dimensional scales from atomic clusters to test coupons to final products
- Net shape, or novel processing of metallic alloys
- Next generation of high temperature alloys
- Surface treatments to enhance structural performance

Electronic and Optical–Photonic Materials

- Deep ultraviolet and electron lithography
- Systems-on-a-chip
- Copper metalization and other interconnects
- Sub-micron plasma processing
- Semiconductor equipment
- Holographic storage materials
- Organic transistors, organic lasers, and LEDs

- Photonic band-gap materials
- Blue-green lasers (gallium nitride materials)
- Semiconductor processing
- Interconnects
- Magnetic storage
- Widegap lasers and display
- Nanomaterials
- Semiconductor Equipment
- Wireless
- Fibers

Superconducting Materials

- High-temperature superconductors
- High-temperature superconductor synthesis
- Processing of highly textured, dense bulk forms for wire and energy storage
- Magnetic phase diagrams and properties
- Statistical mechanical modeling of transport and critical phenomena
- Experimental measurement of flux transport mechanisms
- Modeling of optical and electronic properties
- Physical properties (other than magnetic)
- Development of fluxoid imaging technology
- Thin-film deposition processes
- Epitaxial and patterning techniques

Polymers

- Controlled polymerization
- Metallocene polymerization of olefins
- Living free-radical polymerization
- Atom transfer radical polymerization
- Dendrimer polymerization
- Biologic synthesis
- Supercritical CO_2 as a polymerization medium
- Multicomponent systems
- Blends or alloys
- Block and graft copolymers
- Nanocomposites
- Macrocomposites
- Thin-film laminates
- Interfaces
- Biomedical polymers
- Implants
- Drug delivery
- Electronic–Photonic
- Conducting polymers
- Polymers for display devices

- Resist materials
- Electroluminescent
- Separation media
- Membranes
- Molecular recognition
- Barrier materials
- Modified-atmosphere packaging
- Coatings
- Theory and modeling
- Molecular simulation
- Monte Carlo techniques
- Conformations
- Scaling theory
- Processing
- Rheology
- Flow instabilities
- Computer modeling
- New processes

Catalysts

- Selective oxidation
- Solid acid–base catalysis
- Environmental catalysis
- Catalyst characterization
- Combinatorial catalysis
- Asymmetric catalysts

ATTACHMENT 3

INTERNATIONAL BENCHMARKING OF US IMMUNOLOGY RESEARCH

Panel on International Benchmarking of
US Immunology Research

Committee on Science, Engineering, and Public Policy

This report is dedicated to

Marian (Bunny) Koshland

a pioneer in the field of molecular immunology
and a friend to young immunologists

International Benchmarking of US Immunology Research

Panel Members

IRVING L. WEISSMAN *(Chair)*, Professor of Pathology, Department of Pathology, Stanford University School of Medicine, Palo Alto, California

JAMES ALLISON, Howard Hughes Medical Investigator and Professor, Cancer Research Laboratory, University of California, Berkeley

FREDERICK W. ALT, Howard Hughes Medical Investigator and Professor, Children's Hospital, Harvard Medical School, Cambridge, Massachusetts

HAROLD VON BOEHMER, Professor, Faculte de Medicine Necker – Enfants Malades, Institut Necker, Institut National de la Santé et de la Recherché Medicale, Paris, France

MAX D. COOPER, Howard Hughes Medical Investigator and Professor of Medicine, Pediatrics, Pathology, and Microbiology, University of Alabama, Birmingham

IRWIN FELLER, Director, Institute for Policy Research and Evaluation and Professor of Economics, Pennsylvania State University, University Park

LAURIE H. GLIMCHER, Irene Heinz Given Professor of Immunology, Harvard School of Public Health, and Professor of Medicine, Harvard Medical School, Department of Immunology and Infectious Diseases, Cambridge, Massachusetts

DAVID V. GOEDDEL, President and Chief Executive Officer, Tularik, Inc., South San Francisco, California

HUGH MCDEVITT, Professor of Microbiology and Immunology, Stanford University School of Medicine, Palo Alto, California

DIANE MATHIS, Directeur de Recherches, Institut de Genetique et de Biologie Moleculaire et Cellulaire Institut National de la Santé et de la Recherché Medicale, Strasbourg, France

GUSTAV NOSSAL, Professor Emeritus, Department of Pathology, University of Melbourne, Australia

ROGER M. PERLMUTTER, Senior Vice President, Merck Research Laboratories, Rahway, New Jersey

CRAIG B. THOMPSON, Howard Hughes Medical Investigator and Professor, University of Chicago, Illinois

DON C. WILEY, Howard Hughes Medical Investigator and Professor, Department of Molecular and Cellular Biology, Harvard University, Cambridge, Massachusetts

Project Staff

DEBORAH D. STINE, Study Director
TAMARA ZEMLO, Research Associate
NORMAN GROSSBLATT, Editor

Immunology Benchmarking Guidance Group

PHILIP W. MAJERUS *(Chair)*, Professor of Medicine, Biochemistry, and Molecular Biophysics and Director, Division of Hematology-Oncology, Washington University School of Medicine, St. Louis, MO

MARIAN KOSHLAND (Chair*), Professor of Immunology, Department of Molecular and Cell Biology, University of California, Berkeley

ENRIQUETA C. BOND, President, The Burroughs Wellcome Fund, Durham, NC

RUBY B. HEARN, Senior Vice President, The Robert Wood Johnson Foundation, Princeton, NJ

RICHARD B. JOHNSTON, JR., Department of Pediatrics, National Jewish Medical and Research Center, Denver, CO

DONALD R. MATTISON, Medical Director, March of Dimes Birth Defects Foundation, White Plains, NY

JUNE OSBORN, President, Josiah Macy, Jr. Foundation, New York, NY

**Served from October 1996 to October 1997*

PREFACE

In 1993, the Committee on Science, Engineering, and Public Policy (COSEPUP) of the National Academy of Sciences, the National Academy of Engineering, and the Institute of Medicine issued the report *Science, Technology, and the Federal Government: National Goals for a New Era*. In that report, COSEPUP suggested that the United States adopt the principle of being among the world leaders in all major fields of science so that it can quickly apply and extend advances in science whenever and wherever they occur. The report also recommended that the United States maintain clear leadership in fields that are tied to national objectives, that capture the imagination of society, or that have multiplicative effects on other scientific advances. Those recommendations were reiterated in another Academy report, *Allocating Federal Funds for Science and Technology*, developed by a committee chaired by Frank Press.

Both reports stated that quantitative measures, such as number of dollars spent and number of scientists supported, are inadequate indicators of leadership and that policy decisions about programmatic issues or resource allocation would be better informed by comparative international assessments. To measure international leadership, the reports recommended the establishment of independent panels that would conduct comparative international assessments of scientific accomplishments in particular research fields. COSEPUP indicated that the panels should consist of researchers who work in the specific fields under review (both in the United States and abroad), people who work in closely related fields, and of the research users results who follow the fields closely.

To test the feasibility of the recommendation that panels conduct comparative assessments, COSEPUP has conducted experimental evaluations of three fields: mathematics, materials science and engineering, and immunology. The study panels for the assessments were charged with developing and presenting their findings and conclusions, not recommendations. Specifically, panel members were asked to address the following three questions:

- What is the position of US research in the field relative to the research performed in other regions or countries?
- What key factors influence the US performance in the field?
- On the basis of current trends in the United States and abroad, what will be the future relative position of the United States in the field in the near term and the longer term?

This document presents results of the third and final assessment, that of research in immunology. The panel concluded that the United States is the world leader in immunology, and in its major subfields. In addition, while US dominance is evident in the major sub-fields: cellular immunology, molecular immunology, immunogenetics, and clinical aspects of immunology, and among the world leaders in some parts of subfields, the panel found that US leadership in immunology depends on being able to generate and pursue innovative research ideas. Sufficient funding from both government and private sources, talented researchers, and key infrastructure support mechanisms are instrumental in maintaining US leadership. However, diverse federal and industry priorities, a potential reduction in access to domestic and foreign talent, and the increasing cost of maintaining mice facilities could curtail US ability to capitalize on leadership opportunities in immunology.

Now that all three of the assessments are completed, COSEPUP will begin to discuss the feasibility and utility of the benchmarking process and will make whatever recommendations it deems appropriate.

The committee appreciates all the hard work and dedication of the panel members and thanks them for their help and cooperation in completing this report.

This report has been reviewed by persons chosen for their diverse perspectives and technical expertise, in accordance with procedures approved by the National Research Council's Report Review Committee. The purpose of this independent review is to obtain candid and critical comments that will assist the authors and COSEPUP in making the published report as sound as possible and to ensure that the report meets the institutional standards of objectivity, evidence, and responsiveness to the study charge. The content of the review comments and draft manuscript remains confidential to protect the integrity of the deliberative process. We wish to thank the following for their participation in the review of this report:

Jeffrey Bluestone, Director, Ben May Institute for Cancer Research
Suzan Cozzens, Chair, School of Public Policy, Georgia Institute of Technology
Frank Fitch, Albert D. Lasker Professor Emeritus, Ben May Institute for Cancer Research
Maureen Henderson, Professor Emeritus of Epidemiology and Medicine, University of Washington
Richard Locksley, Department of Medicine, Howard Hughes Medical Institute, University of California San Francisco
Tak Mak, Ontario Cancer Institute, Department of Immunology and Medical Biophysics, University of Toronto
Carl Nathan, Beatrice and Samuel A. Seaver Laboratory, Department of Medicine, Cornell University Medical College
Joseph Newhouse, John D. MacArthur Professor of Health Policy and Management, Harvard University
Philippa Marrack, Howard Hughes Medical Institute, National Jewish Medical and Research Center
Edward Penhoet, Vice Chairman and CEO, Chiron Corporation
Klaus Rajewsky, Institute for Genetics, University of Cologne
Martin Weigert, Department of Molecular Biology, Princeton University
Arthur Weiss, Howard Hughes Medical Institute, University of California

Although those just listed have provided many constructive comments and suggestions, responsibility for the final content of this report rests solely with the author panel and COSEPUP. Finally, the project was aided by the invaluable help of COSEPUP professional staff: Deborah D. Stine, study director, and Tamara Zemlo, research associate.

Phillip A. Griffiths
Chair
Committee on Science, Engineering, and Public Policy

CONTENTS

Tables and Graphs

EXECUTIVE SUMMARY

To be leaders in healthcare and to help maintain a vibrant economy, it is critical that the United States lead the world in immunological research and its clinical applications. The rapid application of immunology's fundamental discoveries has allowed them to contribute to the societal and economic well-being of our country over the past 30 years.

The Committee on Science, Engineering, and Public Policy (COSEPUP) Panel on International Benchmarking of US Immunology Research examined the leadership status of the United States in immunology. The panel was not able to assess this question in an objective way, but it used the expertise and judgment of its members and the limited information available to conclude that the United States is the world leader in immunology.

US dominance is also evident in the major subfields of immunology-cellular immunology, molecular immunology, immunogenetics, and clinical aspects of immunology. This is not a surprising result, given the level of funding of immunology and its consequent effects on the size of the US enterprise, which writes some 60% of the "high impact" (i.e., most cited) immunology papers published each year. However, what is of greater interest, given the size of the enterprise, is that in some parts of subfields international preeminence is more evident.

The panel also found that

- The United States is the world leader in all the major subfields of immunology but is only among the world leaders in some specific sub-subfields.

- Flexibility to pursue original and innovative research ideas has attracted both domestic and international human capital. Federal, state, and private funding have all contributed to a climate ripe for this innovative research.
- Industrial interests have fostered many striking breakthroughs in immunology.
- A scarcity of large-scale clinical trials in immunology can be attributed to shortages in funding and of qualified personnel. In addition, increasing dominance of managed care means that fewer patients are available to academic institutions for clinical trials.
- Shifting federal and industry priorities, a potential reduction in access to domestic and foreign talent, and the increasing cost of maintaining mouse facilities could curtail US ability to capitalize on leadership opportunities.

1

INTRODUCTION

1.1 How Important Is It for the United States To Lead in Immunology Research?

Immunology encompasses fundamental scientific discovery at all levels of biological organization. It is also a practical field that provides, for example, highly specific molecular entities (antibodies) that are the basic tools for identifications (as in diagnosis) and separations in biology, medicine, and industry. Disorders of the immune system are frequent causes of human disease, ranging from congenital and acquired immunodeficiencies, such as AIDS, to autoimmune and inflammatory conditions, such as insulin-dependent diabetes and rheumatoid arthritis. The normal functions of the immune system reject transplanted cells, tissues, and organs. Pharmaceutical and biotechnological interventions to dampen immune responses in autoimmunity, inflammation, and transplantation are important segments of the pharmaceutical industry and clinical medicine, as are attempts to amplify or augment components of the immune system to help eliminate infections, cancers, and parasites that have evaded immunosurveillance.

Institutions that have robust programs of research, training, teaching, and application in immunology have had (and probably will continue to have) opportunities to take advantage of immunology as a science that enriches other biomedical endeavors, to enhance the role of immunology in medicine, and to use immunology in entrepreneurial and industrial efforts. It is therefore a vital US interest to be in a position of leadership in immunology, so that the intellectual, medical, and financial benefits of immunology will be available.

1.2 What Is Immunology?

Immunology has been described as the branch of life sciences that is involved in distinguishing self from nonself. All multicellular (metazoan) organisms are prey to infection or invasion. With the specialization of cells within organisms into organs and tissues that serve distinct functions, subsets of cells that survey other cells or elements for self or nonself markers have evolved to serve immune functions. In humans, as in other vertebrates, the cells that make up the immune system include some that are similar to the innate immune systems of prevertebrates and others—called lymphocytes—that appear to constitute a vertebrate invention, and are responsible for most of the adaptive immune functions of vertebrates.

There are two major classes of lymphocytes: T cells, which develop in the thymus; and B cells, which develop in the bone marrow. Those two classes have different immune functions. For example, in an immune response to a viral infection, B lymphocytes are triggered by the virus to differentiate into mature effector cells that produce and secrete molecules called antibodies. Antibodies can bind to and inactivate or eliminate specific viruses before they enter cells of the body. Viruses can penetrate cells and cause them to use the viral genetic material as templates and instructions to produce more viruses. Once viruses enter the cells, they are largely hidden from circulating antibodies.

To protect the organism against the intracellular phases of viral infection, several populations of lymphocytes, including a subpopulation of inflammatory and killer T cells, collectively isolate and eliminate virus-infected cells. T cells recognize virus-infected cells by means of cell-surface receptors (called T-cell receptors or TCRs) that adhere by molecular complementarity to "flags" on the surface of infected cells. The flags are a class of molecules that make up what is called the major histocompatibility complex (MHC), which picks up degraded fragments (about nine amino acids long) of viral proteins within the cells and bring them to the surface to be detected by TCRs on inflammatory and killer T cells.

Both the magnitude and the quality of the immune responses by these types of T and B cells are regulated by helper T cells. During the course of an infection T and B cells with virus-specific receptors undergo many rounds of cell division; some cell progeny are destined to be immediate effectors of the response, and others are retained as "memory" cells. Long after an infection (or vaccination), the expanded number of memory cells guarantees that a second exposure to the same virus will be met by an expanded response which develops more rapidly, providing effective immunity before serious consequences of the infection develop. During development of the T and B lymphocytes from their precursors, members of the population that have receptors to self are usually eliminated, inactivated, or not expanded. Thus, the adaptive immune system usually ignores self and responds to nonself by providing early and effective immunity and lifelong immune memory. Much is known about this complex system, but much is yet to be learned, so immunology is still an attractive subject for training and research. For example, many immune diseases are known to be the result of mutations or alterations in particular components of the system, whereas others have an unknown etiology.

Immunology attracts diverse life scientists. Perhaps because the cells of the immune system are easily obtained, the system has often been used as a leading-edge subject for studies in other disciplines. Study of homogeneous lymphocyte populations, for example, leads to research in many aspects of signal transduction, wherein cell-surface receptor engagement signals cells to divide, differentiate, or die. It can be argued that we know more about vertebrate developmental immunology than about any other developmental system, including the first isolated stem cell in any system. Much of what is known about cell-surface adhesion and recognition receptors, the genes that encode them, and the evolution of these genes comes from studies of cells of the immune system. That cells can communicate by secreted protein messages called cytokines was elucidated largely through study of cells and cytokines of the immune system.

1.3 Immunology As an Academic Discipline

Immunologists are at work in virtually every life science department or division. But there are very few departments of immunology in academe. The multidisciplinary nature of immunology research is probably a major reason that immunology is so well connected with the more traditional subjects (such as biochemistry, genetics, and microbiology), whose approaches define their disciplines. It also explains the ready translation of discoveries in immunology to such clinical subjects as rheumatology, surgery (in transplantation), endocrinology (in diabetes), neurology (in multiple sclerosis), and allergy.

Although those connections have served immunology and the other subjects well and have probably protected immunologists from the isolation that their jargon could lead them into, immunology might be less of a force in academic politics and less of a presence in the undergraduate and graduate curricula of many universities than it would otherwise be because it does not have departmental status at most institutions. Not being or being in a discipline with departmental status, immunology and immunologists are unevenly distributed in the totality of US academic institutions. Entrepreneurial and clinical efforts in immunology largely have the same uneven distribution.

1.4 What Is the International Nature of Immunology?

Excellent research in immunology is conducted throughout the world. Researchers are part of a tightly knit and highly collaborative international community and hence, immunology as a discipline has become an international effort. International collaboration in the different subfields of immunology has facilitated exchanges in information that have enabled exciting breakthroughs to be made. Factors that have contributed to international collaboration have been the training of young scientists from around the world in graduate institutions in the United States, training of young US scientists in foreign immunology centers, internationally attended scientific conferences, the increasing facility of electronic forms of communication, and the use of English as the standard tongue of communication.

1.5 What Are Some Caveats?

Immunology is an essentially multidisciplinary field, and immunological research overlaps with many other disciplines, including molecular and cellular biology, genetics, and biochemistry. Immunology serves as a foundation for the design and testing of varied biologic hypotheses. Therefore, this benchmarking assessment will be valuable not only to the field of immunology, but also to other biological disciplines. Conversely, although this is a definitive strength of this report, it must be noted that it was sometimes difficult to identify and characterize specific attributes that apply solely to immunology.

Additional caveats apply to the method used in this analysis. Given the lack of quantitative data that can be compared on an international basis, the panel used a number of techniques and, looked at the degree to which the results conformed to develop its conclusions. The panel was not able to assess this question in an objective way, but it used the expertise and judgment of its members and the limited information and data available to develop its conclusions. More details on the methods and the limitations of each are provided in Chapter 2.

1.6 Panel Charge and Rationale

The panel was asked to conduct a comparative international assessment to answer three questions:

- What is the position of the US research in the field relative to the research performed in other regions or countries?
- What key factors influence the US performance in the field?
- On the basis of current trends in the United States and abroad, what will be the future relative position of the United States in the field in the near term and the longer term?

The panel was asked only to develop findings and conclusions not recommendations. Its primary objective was to obtain a comprehensive overview of the field of immunology that included characterizing the key factors of the field, assessing the resources necessary for conducting and supporting immunologic research, and identifying trends in the types of research being done in the field. The panel strove to maintain an international perspective as it collected and analyzed the data for this report.

The panel assessed the current position of the United States relative to leadership in four subfields of immunology, and the benchmarking results themselves are presented in Chapter 2 of this report. The determinants of leadership that have influenced US advancement in the field are discussed in Chapter 3. Chapter 4 assimilates past leadership determinants and current benchmarking results to predict future US leadership status in the field.

2

BENCHMARKING RESULTS

2.1 Methods

The assessment of a country's relative standing in research performance is subject to multiple complexities. Many measures could be used to evaluate the position of US-based research in immunology, but each suffers from generic or specific limitations. Such limitations include the collaborative and international scope of immunological research, which makes the drawing of national lines somewhat arbitrary; the inequality inherent in comparing a large, common enterprise with multiple smaller ones; and the difficulties of sifting information on the specific field of immunology from that on related research fields in large, aggregate databases. In addition, in the case of this panel's operations, budget and time constraints effectively ruled out any major undertakings to generate new data sets.

Within those constraints, the panel's strategy was to use three mainstream performance measures: reputation survey, citation analysis, and journal publication analysis, it then relied on the convergence of findings to compensate for the specific shortcomings of each measure. Although these measures have some independence, the degree of independence could not be assessed because the panel could not compensate for all of the shortcomings of the methodology.

As shown in Figures 2.1 and 2.2, the United States is clearly dominant in number of papers published in immunology. It produced 63% of the world's high impact (i.e., most cited) papers in immunology in 1981-1996; no other country produced 10%, and only one produced even 5%. Thus, comparisons are made in terms of the United States versus the rest of the world instead of on a country by country basis.

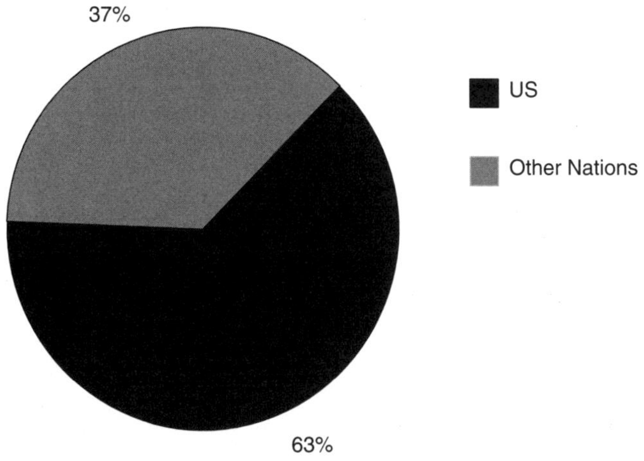

FIGURE 2.1 Contribution of United States and other nations to high-impact immunology papers in 1981–1996. Source: Calculated from data obtained from the Institute for Scientific Information database on high-impact papers in immunology.

2.1.1 Reputation Survey

To estimate reputation, the panel conceived a virtual congress in which leading experts in immunology were asked to identify potential participants. Specifically, the field was divided into four major subfields-cellular immunology, molecular immunology, immunogenetics, and clinical aspects of immunology—and each was subdivided into four to 10 sub-subfields. The panel (composed of 11 US-based and three non-US-based scientists) then identified five to 15 respected leaders in each sub-subfield, based in the United States and in other countries, and polled each in person, by telephone, or by mail. The pollees were asked to imagine that they were about to organize an international congress session on their particular sub-subfield and to furnish a list of five to 20 potential speakers.

When the initial lists were considered, it became clear that there was a bias related to the laboratory location of the pollees: US-based investigators routinely named a higher percentage of Americans than did non-US-based investigators. The nationality of the poller also appeared to have an influence: the three non-US pollers often obtained a list more enriched in non-US speakers. Thus, additional subfield leaders were polled to approximate a 50:50 US: non-US ratio.

An advantage of our approach is that it incorporates the opinions of a variety of respected members of the immunology community: up-and-coming leaders as well as established ones, investigators from all over the world, and leaders of all sub-subfields, both basic and clinical. The

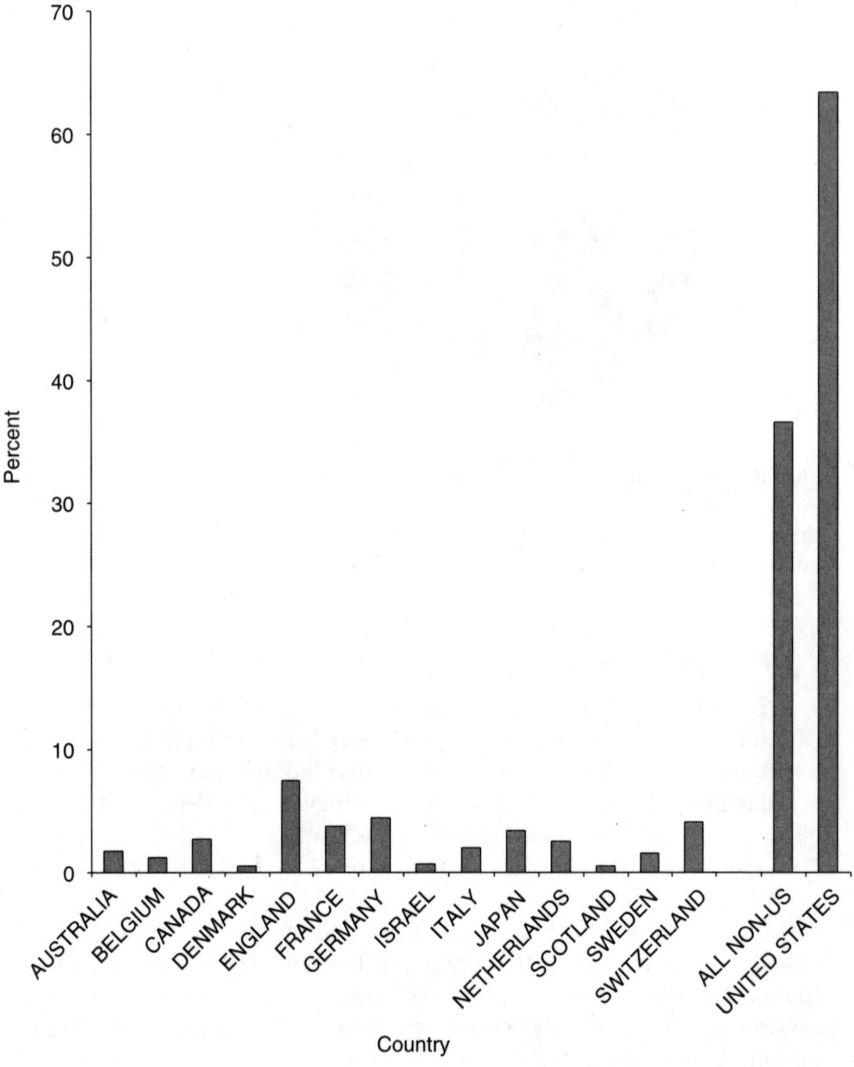

FIGURE 2.2 Percentage of world's high-impact papers in immunology from 1981–1996, by country. Source: Calculated from data obtained from the Institute for Scientific Information database on high-impact papers in immunology.

disadvantages are multiple: variable modes of polling, a variable sample size that was usually too small to allow statistical treatment of the data, lack of objective criteria, and nonproportionate sampling of researchers in countries outside the United States.

2.1.2 Citation Analysis

To identify the most frequently cited authors of immunologic research articles, a "high-impact" immunology database was commissioned from the Institute for Scientific Information (ISI). (See *http://www.isnet.com/products/rsg/impact.html* for more information on "high-impact" paper methodology.) The ISI database was scanned over the years 1990-1997. For each year, the 200 most-cited papers in journals relevant to the field of immunology were listed. The authors having more than five papers on the list were ranked according to average number of citations per paper (174 authors ranging from 70.2 to 638.5/paper), and the country of the laboratory of each was listed. In addition, panel staff determined whether each author was also cited on the virtual-congress lists.

The contribution of the United States and other nations to immunology citations from 1981-1997 is shown in Figure 2.3. Of the 174 authors identified in the scan of the ISI database, 72% including the top 112 authors identified were in US-based laboratories. The ISI database indicates that US laboratories produced 63% of the papers, which garnered 66% of the citations, in journals relevant to immunology during the period 1981-1997.

The major strengths of this mode of analysis are its relative objectivity and its providing a basis of comparison with the virtual-congress polling data. However, the analysis suffers from some flaws related to

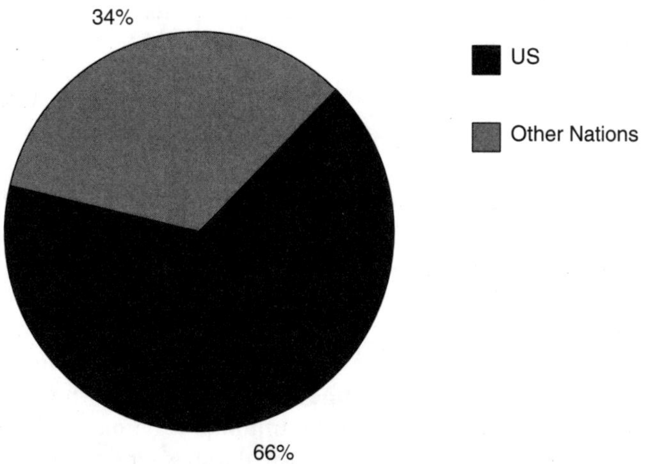

FIGURE 2.3 Contribution of United States and other nations to high-impact immunology citations in 1981-1997. Source: Calculated from data obtained from the Institute for Scientific Information database on high-impact papers in immunology.

the organization of the ISI database. Data from several of the top-level general journals (such as *Cell*) and immunology journals (such as *Journal of Experimental Medicine*) were not included. In addition, the list of authors is truncated after the first 15 names, possibly excluding authors who participated in large clinical trials. An additional limitation imposed by the database was that there was no breakdown according to subfield and sub-subfield of immunology.

2.1.3 Journal Publication Analysis

The panel identified four leading general journals (*Science*, *Cell*, *Nature*, and *Blood*) and one of the top journals focused specifically on immunology (*Immunity*). Panel members scanned the tables of contents of each of the journals for 1995-1997, identifying immunology papers in the general journals and the laboratory nationality of the principal investigator and the subfield in all the journals. In addition, a small sample from the *Journal of Experimental Medicine* was analyzed.

Additional journals—such as *Nature Medicine, Lancet, Journal of Clinical Investigation*, and *New England Journal of Medicine*—could have been analyzed as well but such an analysis was beyond the resources of this panel. Others—such as the *European Journal of Immunology, Journal of Immunology* (US), *International Immunology* (Japan), and *Immunology* (UK)—reflect mainly the country of research origin and so are not appropriate here.

This approach appears to be procedurally the least biased of the three. It does suffer from the usual weaknesses of any analysis based on publication data, including the dominance of English as the language of international discourse and the vagaries of the peer-review journal-acceptance process.

2.2 Results

Although the three criteria for evaluating US research efforts in immunology were quite distinct and had different strengths and flaws they led to basically the same conclusion: immunology research in the United States is pre-eminent in the world. The data supporting that conclusion are summarized below and in the following tables.

2.2.1 Reputation Survey

The data-collection measures for the different sub-subfields proved highly variable, including the number of pollees (3-17), the fraction of US-based pollees (41% - 83%), and the number of names cited per pollee (a few to more than 20). In addition, the database for the individual sub-subfields was far too small to permit any kind of statistical analysis. Those points are partially illustrated by some of the entries in Table 2.1. The panel decided to emphasize the data pooled by subfields in which there was generally much less variation in the data-collection measures, although attention was still paid to findings in the context of the sub-subfield groupings for any nuances of interest.

In all four subfields, a clear majority of the names cited were investigators directing US laboratories 60 - 70% in all cases. However, the position of leadership was not so evident in some domains, for example:

- In cellular immunology, the sub-subfields of lymphocyte development and self-nonself recognition.
- In molecular immunology, the sub-subfield of NK receptors.
- In immunogenetics, the sub-subfield of inherited immunodeficiency.
- In clinical aspects of immunology, the sub-subfields of tumor immunology and transplantation and immunosuppressive drugs.

In those domains, the proportion of US-based investigators was closer to 50%. The statistical significance of the differences could not be tested, but in general they correspond well to the collective opinion of the panel numbers.

2.2.2 Citation Analysis

The results of the panel's citation analyses are shown in Figure 2.4 and Table 2.2. As shown in Figure 2.4, the top 3 countries for the immunology field based on percentage of the world's citations from 1981-1997 in immunology are:

1. United States
2. England
3. Switzerland

Another measure that can be used is relative citation impact (RCI). RCI is the country's share of the world's citations in the field, divided by its share of world publications in the field. It can be thought of as a comparison of a country's citation rate for a particular field with the world's citation rate for the field. A relative citation impact greater than 1 shows that the country's rate for the field is higher than the world's. Some believe RCI is a measure of both the influence and the visibility of a country's research (as disseminated through publications) and it gives some indication of the quality of the average paper. As shown in Table 2.2, the top 3 countries based on relative citation index for 1981-1997 are:

1. United States
2. Belgium
3. Australia

TABLE 2.1 Immunology International Reputation Survey Results

		Pollees					Virtual-Congress Speakers				
Subfield	Sub-subfield	No. of Pollees	No. US	% US	No. Non-US	% Non-US	No.	No. US	US	No. Non-US	% Non-US
Cellular immunology and immunology	1. Lymphocyte development	17	7	41.2	10	58.8	189	98	51.9%	89	47.1%
	2. Cellular interactions	10	6	60	4	40	62	46	74.2%	16	25.8%
	3. Self-nonself recognition	11	6	54.5	5	45.5	122	59	48.4%	62	50.8%
	4. Cellular traffic adhesion	4	3	75	1	25	75	47	62.7%	28	37.3%
	5. Host-parasite interactions	x	x	x	x	x	x	x	x	x	x
	6. Innate immunity	5	3	60	2	40	67	49	73.1%	18	26.9%
	7. Mucosal immunity	4	3	75	1	25	99	70	70.7%	29	29.3%
Molecular aspects of immunology	1. B- and T-cell genes	8	3	37.5	5	62.5	113	65	57.5%	47	41.6%
	2. NK receptors	11	6	54.5	5	45.5	124	76	61.3%	48	38.7%
	3. Receptors and tolerance	8	6	75	2	25	71	51	71.8%	20	28.2%
	4. Transcription	8	4	50	4	50	88	72	81.8%	16	18.2%
	5. Antigen uptake	5	4	80	1	20	79	46	58.2%	32	40.5%
	6. Structural analysis	7	5	71.4	2	28.6	34	36	76.5%	8	23.5%
	7. Apoptosis	4	3	75	1	25	28	22	78.6%	6	21.4%
Immunogenetics	1. Evolution of MHC	x	x	x	x	x	x	x	x	x	x
	2. Autoimmune disease	4	2	50	2	50	63	41	65.1%	22	34.9%
	3. Inherited immunodeficiency	10	6	60	4	40	145	74	51.0%	71	49.0%
	4. Neoplastic immune disorders	6	5	83.3	1	16.7	61	41	67.2%	20	32.8%
Clinical aspects of immunology	1. Immunopathologies and immune interventions	13	8	61.5	5	38.5	258	157	60.9%	99	38.4%
	2. Tumor immunology	4	2	50	2	50	34	26	76.5%	8	23.5%
	3. Transplantation and immunosuppressive drugs	13	7	53.8	6	46.2	189	108	57.1%	81	42.9%

Source: Nonrandom polling data obtained by panel members interviewing equal numbers of US and non-US immunologists whose research the panel members felt had made significant contributions to specific subfields of immunology. The chosen immunologists (pollees) were instructed to select other prominent immunologists in their subfields that they would invite to talk at a virtual immunology conference. Invited researchers are referred to here as virtual congress speakers. Both the pollees and virtual congress speakers remain anonymous and are reported as frequencies and percentages of US and non-US researchers in table.

Note: Percentages might not add to 100% because of rounding errors or missing data; x = data insufficient.

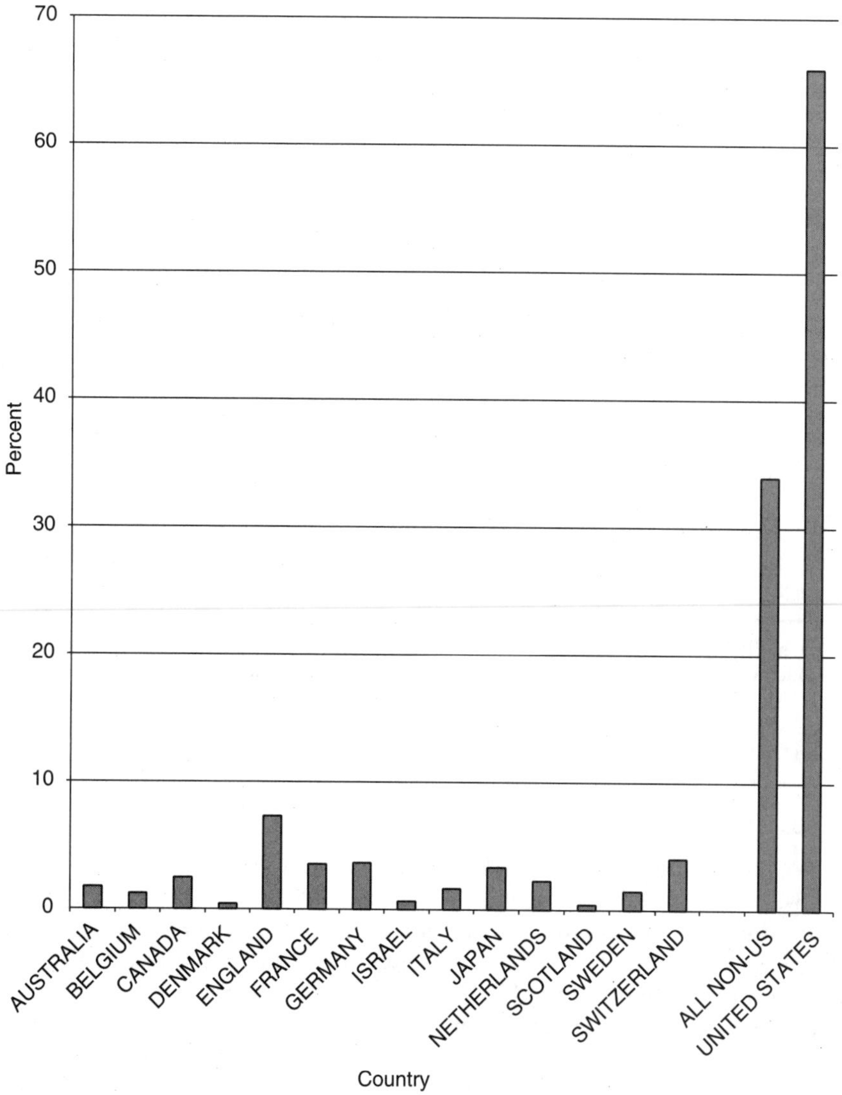

FIGURE 2.4 Percentage of world's citations to high-impact papers in immunology in 1981-1997, by country. Source: Calculated from data obtained from the Institute for Scientific Information database on high-impact papers in immunology.

A bit disconcertingly, only 54% of the authors identified in the ISI Immunology high-impact citations were also cited in the survey conducted by the panel. The discrepancy can probably be attributed to flaws in the two approaches, in particular such defects in the ISI database as the exclusion of critical journals such as *Cell* and the *Journal of Experimental Medicine*.

TABLE 2.2 Relative citation impact of high-impact papers in immunology, by country, 1981-1997

Country	No. Citations	Fraction of World's Citations	No. Papers	Fraction of World's Papers	RCI[a]
United States	632,326	66.03	2,360	63.39	1.04
Belgium	11,626	1.21	44	1.18	1.03
Australia	16,701	1.74	65	1.75	1.00
Switzerland	38,487	4.02	153	4.11	0.98
England	69,645	7.27	278	7.47	0.97
Japan	31,689	3.31	127	3.41	0.97
France	33,877	3.54	139	3.73	0.95
Sweden	13,865	1.45	57	1.53	0.95
Canada	23,329	2.44	101	2.71	0.90
Israel	6,098	0.64	27	0.73	0.88
Netherlands	21,378	2.23	95	2.55	0.87
Denmark	3,910	0.41	18	0.48	0.84
Germany[b]	34,927	3.65	164	4.41	0.83
Italy	15,746	1.64	75	2.01	0.82
Scotland	4,009	0.42	20	0.54	0.78
Totals	957,613	—	3,723	—	—

Notes: [a]RCI is the relative citation impact, a country's percentage of world citations divided by its percentage of world papers. Threshold set at minimum of 17 papers.
Source: Calculated from data obtained from the Institute for Scientific Information database on high-impact papers in immunology.

2.2.3 Journal Publication Analysis

The results of the journal publication analysis are shown in Tables 2.3-2.7. US-based investigators produced about three-fourths of the immunology papers published in the journals *Cell, Science*, and *Immunity* in 1995-1997. US-based researchers contributed about two-thirds of the immunology papers published in *Nature* in 1995-1997. In contrast, scientists in countries other than the United States published more immunology articles in *Blood* than US-based researchers in those years (54% and 46%, respectively). The immunological subfield in which an international presence was especially strong was immunogenetics for the papers published in *Cell* and *Blood*. However, in *Nature*, *Science*, and *Immunity*, US-based researchers published more papers in immunogenetics than non-US-based researchers. The United States had more papers published in all five journals in the subfield of molecular immunology. The subfield of cellular immunology and clinical aspects were dominated by US-based researchers in all journals examined except *Blood*. Because of its extensiveness, only 6 months of the *Journal of Experimental Medicine* was analyzed, as shown in Table 2.8.

TABLE 2.3 Authorship of Immunology Papers in *Blood*, 1995-1997

Subfield	No.(%) US	No.(%) Non US	Total
Cellular immunology	58(46.4)	67(53.6)	125
Molecular immunology	47(54.7)	39(45.3)	86
Immunogenetics	10(37.0)	17(63.0)	27
Clinical aspects	158(44.5)	197(55.5)	355
TOTALS	273(46.0)	320(54.0)	593

Source: Original analysis conducted for this report. Panel members reviewed tables of contents and decided subfields of immunology of each paper.

TABLE 2.4 Authorship of Immunology Papers in *Cell*, 1995-1997

Subfield	No.(%) U.S.	No.(%) Non U.S.	Total
Cellular immunology	14(60.9)	9(39.1)	23
Molecular immunology	45(75.0)	15(25.0)	60
Immunogenetics	0(0)	1(100)	1
Clinical aspects	12(75.0)	4(25.0)	16
TOTALS	71(71.0)	29(29.0)	100

Source: Original analysis conducted for this report. Panel members reviewed tables of contents and decided subfields of immunology of each paper.

TABLE 2.5 Authorship of Immunology Papers in *Immunity*, 1995-1997

Subfield	No.(%) U.S.	No.(%) Non U.S.	Total
Cellular immunology	113(72.9)	42(27.1)	155
Molecular immunology	145(78.3)	40(21.6)	185
Immunogenetics	32(72.7)	12(27.2)	44
Clinical aspects	26(81.2)	6(18.7)	32
TOTALS	316(76.0)	100(24.0)	416

Source: Original analysis conducted for this report. Panel members reviewed tables of contents and decided subfields of immunology of each paper.

TABLE 2.6 Authorship of Immunology Papers in *Nature*, 1995-1997

Subfield	No.(%) U.S.	No.(%) Non U.S.	Total
Cellular immunology	37(68.5)	17(31.5)	54
Molecular immunology	48(61.5)	30(38.5)	78
Immunogenetics	6(66.7)	3(33.3)	9
Clinical aspects	9(64.3)	5(35.7)	14
TOTALS	100(64.5)	55(35.5)	155

Source: Original analysis conducted for this report. Panel members reviewed tables of contents and decided subfields of immunology of each paper.

TABLE 2.7 Authorship of Immunology Papers in *Science*, 1995-1997

Subfield	No.(%) U.S.	No.(%) Non U.S.	Total
Cellular immunology	44(67.7)	21(32.3)	65
Molecular immunology	68(82.9)	14(17.1)	82
Immunogenetics	6(60.0)	4(40.0)	10
Clinical aspects	46(66.7)	23(33.3)	69
TOTALS	164(72.6)	62(27.4)	226

Source: Original analysis conducted for this report. Panel members reviewed tables of contents and decided subfields of immunology of each paper.

TABLE 2.8 Authorship of Immunology Papers in the *Journal of Experimental Medicine*, February 1996-July 1996

Subfield	No.(%) U.S.	No.(%) Non U.S.	Total
Cellular immunology	53(58.9)	37(41.1)	90
Molecular immunology	61(77.2)	18(22.8)	79
Immunogenetics	13(86.7)	2(13.3)	15
Clinical aspects	37(64.9)	20(35.1)	57
TOTALS	164(68.0)	77(32.0)	241

Source: Original analysis conducted for this report. Panel members reviewed tables of contents and decided subfields of immunology of each paper.

2.3 Summary

According to all three evaluation methods, US-based research in immunology plays a dominant role in the worldwide effort. Within the limitations of each measure and the limits of the panel's use of them, the data produce a strikingly consistent outcome: all three approaches assigned a 2:1 to 3:1 dominance vis-a-vis the rest of the world. Of course the dominance needs to be weighed in relation to the relative richness of the United States in numbers of investigators, institutions and resources.

3

KEY FACTORS

The panel identified five key factors that influence the international leadership status of US immunology research. These factors are

- Funding
- Human Resources
- Infrastructure
- Biotechnology and Pharmaceutical Firms
- Clinical Trials

Each is assessed in more depth below relative to how the United States compares to other countries.

3.1 Funding

Both the US and foreign members of the panel generally agreed that the structure and financial-support mechanisms of the major research institutions in the United States and the structure and mechanisms for provision of research-grant support by government and private granting agencies constitutes a major factor in the success of the US scientific enterprise in immunology and in almost every field of biomedical research.

The reasons have to do with the organization of higher education and research in contrast with the situation in the United States. Many foreign countries, US universities, medical schools, and research institutes are either privately supported or supported by individual state governments-separate administrative units, under the federal system in the United States. Thus, there is a great diversity of private institutions and a great diversity of mechanisms and approaches for funding state-supported institutions and philanthropically supported institutions throughout the United States. Most important the management, regulation, and governance of private institutions are determined by the institutions, and are therefore somewhat removed from the direct effects of federal funding decisions and federal granting agencies.

In contrast, in Europe, Japan, Australia, and several other countries, the central government supports research institutions, universities, and medical schools and allocates research-grant support for specific research projects of specific people. Government regulation of hiring, personnel practices, and many other aspects of operating research laboratories are therefore centrally controlled and do not permit the diversity that is characteristic of the US scientific enterprise. In addition to the diversity of US institutional organization and support, many research programs throughout the United States have enjoyed a greater degree of support from pharmaceutical and biotechnology firms than is true in Europe.

A second major factor in fostering innovation, creativity and rapid development of new technologies is the National Institutes of Health (NIH) model of research-grant allocation and funding: almost all research (except small projects funded by contracts) is initiated by individual investigators, and the decision as to merit is made by a dual-review system of detailed peer review by experts in each subfield of biomedical science.

In this system, a grant is given to an individual investigator, essentially regardless of the investigator's academic rank or position, as long as he or she is given principal investigator status by his or her institution. Almost all institutions grant principal investigator status to scientists at the beginning of their independent careers, almost always after completion of a postdoctoral fellowship. Individual investigators in universities, medical schools, and research institutions are thus empowered to be individual entrepreneurs. They are not subject to any type of review or control of their chosen research subjects by department chairs, other faculty colleagues, or other scientific colleagues in their institutions. This system has prevented the development of hierarchical research groups of the sort that are seen in many other countries, and it has fostered innovation and independent research initiatives to an amazing degree.

Another major source of funding of immunology (and other biomedical subjects) is the Howard Hughes Medical Institute (HHMI). HHMI selects and retains investigators (rather than projects) largely on the basis of their track record. HHMI-selected investigators are widely regarded as among the most distinguished and productive in the field at both the senior and junior investigator levels. A key to the process has been the selection of external reviewers solely on the basis of their scientific accomplishments and their standing in the field. HHMI provides superb infrastructure for its scientists, who are staff members of HHMI, but whose laboratories are integrated into major academic and research institutions, mainly in the United States. The HHMI scientists are much freer to follow their imaginations and to change the course of their projects than NIH funded investigators, in that the principal evaluation of HHMI investigators is based on productivity, whereas NIH evaluates progress mainly on prescribed projects. The funding of HHMI investigators has substantially enhanced their productivity and has relieved the pressure on NIH to fund meritorious other projects. Additional private sources of immunology funding in the United States are the American Cancer Society, the Juvenile Diabetes Foundation, the Arthritis Foundation, and the Multiple Sclerosis Society.

Funding for training grants for predoctoral fellows and postdoctoral fellows also comes from a wide variety of institutes of NIH and from private sources. Both types of funding have, over the last 40 years, influenced how science is organized in the United States. There are two major results of this entrepreneurial, individual-based system:

- It has led to the development of multiple centers of excellence in immunology and many other fields of biomedical research at many centers around the country.
- Many key research centers are based in or closely attached to large medical centers. This stimulates the expansion and application of immunology to many clinical problems and the study of many problems in basic immunology.

Because immunology research in the United States is based largely in medical institutions and because research, training, and clinical activities go on in parallel in these institutions, interdisciplinary research, development of clinical applications, and the application of basic immunology in solving clinical problems have all been fostered.

Further, NIH and several private funding agencies foster basic scientific training for clinically trained people. Many medical schools have people in their departments of medicine, pediatrics, and surgery with both a full clinical background and a basic-research background in immunology and related fields.

The current apparent eminence of US-based immunologists should not be taken as leadership in all aspects of training and immunology research, however. As shown in Table 3.1, important research in immunology rewarded by Nobel Prizes has been carried out by 16 laureates, 12 of whom were not US citizens (though some now conduct their research in the United States).

TABLE 3.1 Analysis of Nobel Prizes Presented for Immunology Research

Prize	Laureate	Citizenship	Research Done In	Currently
1951	Max Theiler	South Africa	South Africa	—
1957	Daniel Bovet	Switzerland	Switzerland	—
1960	F. Macfarlane Burnet	Australia	Australia	—
	Peter Medawar	Great Britain	Great Britain	—
1972	Rodney R. Porter	Great Britain	Great Britain	—
	Gerald M. Edelman	United States	United States	United States
1977	Rosalyn S. Yalow	United States	United States	United States
1980	George D. Snell	United States	United States	—
	Jean Dausset	France	France	France
	Baruj Benecerraf	United States	United States	United States
1984	Cesar Milstein	Great Britain	Great Britain	Great Britain
	Georges J.F. Kohler	Germany	Switzerland	—
	Niels K. Jerne	Denmark	Switzerland	—
1987	Susumu Tonegawa	Japan	Switzerland/ United States	United States
1996	Peter C. Doherty	Australia	Australia/ United States	United States
	Rolf M. Zinkernagel	Switzerland	Australia/ United States/ Switzerland	Switzerland

Source: Analysis conducted by panel members for this report.

3.2 Human Resources

Its flexibility, diversity, and freedom to originate new approaches has made the United States a very attractive environment for talented researchers from other countries. This has given US research institutions a greater ability than foreign institutions to attract graduate students and postdoctoral fellows from other countries.

The flexibility of funding based primarily on peer review and the merit of applications have made the United States a more attractive country for talented researchers at higher ranks to settle and pursue their research careers. There is a much greater flow of foreign researchers into the United States than the opposite direction because of the lack of barriers (other than language) in the United States.

The US secondary-education system has numerous deficiencies. However, the flexibility allows students, particularly talented students, to obtain research experience in their own institutions and through summer programs, such as those at the Jackson Laboratories and the Cold Spring Harbor Laboratories. Despite those excellent opportunities at the predoctoral level for a small subset of students, the percentage of doctorate recipients with US citizenship in the combined fields of immunology, microbiology, and virology has decreased from 88.0% in 1980 to 77.9% in 1995. This drop of 11.5% in the proportion of recipients with US citizenship is not as steep as the drop of 22.1% in all the life sciences combined (82.4% in 1980 to 64.2% in 1995). The percentage of foreign doctorate recipients in immunology that were planning to obtain postdoctoral fellowships has increased from 7.0% in 1976-1985 to 13.0% in 1986-1996[1].

3.3 Infrastructure

The United States has been fortunate in the development of mouse genetics, inbred strains of mice, and many other variations of the basic inbred strains that have been fostered and developed at the Jackson Laboratories. A result has been that a much higher percentage of US immunology research is carried out on the laboratory mouse than was initially true in Europe and Asian countries.

The capital investment by the NIH, the National Science Foundation, and private research-granting agencies in infrastructure, equipment, and buildings for research has been a major source of growth in immunology and many other fields of biomedical research in the United States.

But European countries have proved more adept at large-scale clinical research projects in immunology than the United States, where the great diversity of institutions and institutional support has balkanized the research effort. This works to the detriment of efficient, large-scale clinical research in the United States, once basic research has led to the development of new therapeutic approaches. The European adeptness is due to many factors. In some cases, it is because of the centralized government control of medical schools and research institutions. In others, it is because physicians are able to maintain a single life-long comprehensive record of patients, which makes it easier to randomize individual patients or practice. Furthermore, in some countries, such as the United Kingdom, clinical-trial methodology has been a special interest of the medical research council and by a national policy that uses randomized trials as a way to introduce new treatment or diagnostic tests.

3.4 Biotechnology and Pharmaceutical Firms

Because of the nature of the venture-capital industry in the United States, the greater flexibility of this industry, and its willingness to fund

[1] All data in this paragraph is from special analysis conducted by NRC Office of Scientific and Engineering Personnel of Survey of Earned Doctorates database for this study.

small biotechnology startup firms, particularly those involved in molecular biology and recombinant-DNA technology, there has been a remarkable growth in biotechnology and a gradual shift of those firms into large pharmaceutical firms. In the last 7 years, although the number of biotechnology companies worldwide has been rather static at approximately 1,275, the amount of money spent on research and development by the industry has almost doubled from $4.9 billion to $9.9 billion (Ernst & Young, 1998a). The result of this phenomenal growth has been the creation of a new source of employment for PhD and MD trainees in immunology, which has attracted many graduate students into immunology.

Industrywide data on the amount of money spent on immunology-specific research are not available, so the panel chose to examine trends in industry-supported research for the entire biotechnology industry. Biotechnology industry support for research is much greater in the United States than in Europe as shown in Tables 3.2-3.3 (Ernst & Young, 1998b). This financing of research and the use of many academic researchers for consultation in biotech firms and large pharmaceutical firms have provided relatively direct avenues for postdoctoral immunologists to obtain employment, to move across disciplines, and to capitalize rapidly on technology developments that are fostered primarily in biotech firms. In addition, the role of many US academic researchers in founding or participating in the founding of biotechnology firms has enhanced the linkage between academic and industrial research in immunology. In some cases (decidedly a minority), the necessity for patent protection has sometimes impeded the flow of information from research developments in biotechnology and pharmaceutical firms. The ability of biotech firms and large pharmaceutical firms to take discoveries from academic research into startup companies and then large firms and into clinical application has been an overall benefit for the development of clinical immunology in the United States. This entrepreneurial approach has also translated into an economic advantage for the United States over other countries. As shown in Figure 3.1, the United States has a net positive trade balance in biotechnology-based products that was in the low $600 million range in 1990, rose to almost $1 billion in 1994 and then decreased to about $650 million in 1996. (NSF, 1998: Appendix Table 6-6)

3.5 The Clinical Trial

There is a shortage of people in the United States trained to design and administer large-scale trials of new immunology-based therapies. In addition, the impact of managed care has narrowed the patient base available for this type of clinical research, except in large, nonprofit managed-care organizations, such as the Kaiser-Permanente organization.

TABLE 3.2 Biotechnology Industry Comparable Metrics (Ecu in Millions)

European Companies

Biotech Company	Market Cap	Turnover	Profit/ Loss	R&D Costs	Employees
British Biotech	1,015	29.1	–42.6	54.2	454
Qiagen	699	68.2	8.1	7.6	650
Innogenetics	552	25.9	–0.9	12.6	380
Biocompatibles International	544	16.4	–27.6	12.2	393
Shire Pharmaceuticals	531	34.4	–0.2	16.1	390
Cortecs	412	11.5	–17.5	17.2	258
Genset	392	14.9	–17.1	23.1	355
Chiroscience	358	17.2	–27.9	33.2	320
NeuroSearch	357	2.7	–8.9	10.1	110
Celltech Group	340	6.4	–17.9	31.2	220
Scotia Holding	309	42.1	–30.8	34.3	420

Multinational Company	Market Cap	Turnover	Profit/ Loss	R&D Costs	Employees
Novartis	99,554	19,590.4	3,274	2,320.3	87,239
Glaxo Welcome	76,153	11,915.7	4,011	1,714.2	52.501
Smithkline Beecham	52,447	11,639.5	2,464	1,255.8	55,400
Zeneca	303,312	7,749	1,120	975.1	31,100
Astra	25,425	5,074.2	1,153	988.3	22,206
Baver Group	24,741	27,912	1,498	2,011.7	144,600
Hoechst Marion Roussel	18,590	7,091	838	1,206	40,500

US Companies

Biotech Company	Market Cap	Turnover	Profit/ Loss	R&D Costs	Employees
Amgen	14,144	2,115.8	624.7	485.1	4,646
Chiron	3,588	1,206.3	50.5	340.8	7,434
Genentech	2,469	888.4	108.4	432.7	3,071
Biogen	2,300	254.5	37.7	121.3	675
Alza	2,264	428.1	84.5	130.5	1,652
Genzyme	1,938	488.7	–67.1	194.8	3,516
Immunex	1,319	140.6	–49.6	89.1	808

Multinational Company	Market Cap	Turnover	Profit/ Loss	R&D Costs	Employees
Merck	114,894	18,216.9	3,566	1,366.1	49,100
Johnson & Johnson	78,847	19,531.6	2,652	1,750.1	89,300
Bristol-Myers Squibb	74,174	13,840.2	2,618	1,172.3	51,200
Eli Lilly	55,517	6,749.7	1,400	1,093.3	29,200
Pfizer	35,479	10,386.8	1,772	1,547.1	46,500

Source: Ernst & Young, 1998a.

TABLE 3.3 Entrepreneurial Life Science Highlights (Ecu in Millions)

Europe

	Public Companies			Industry Total		
	Current Year	Prior Year	Percent Change	Current Year	Prior Year	Percent Change
Financial						
Revenues	648	433	50%	2,725	1,721	58%
R&D expense	534	243	120%	1,910	1,508	27%
Net loss	347	73	375%	2,020	1,113	81%
Industry						
Number of Companies	61	49	24%	1,036	716	45%
Employees	8,418	5,315	58%	39,045	27,500	42%

USA

	Public Companies			Industry Total		
	Current Year	Prior Year	Percent Change	Current Year	Prior Year	Percent Change
Financial						
Revenues	12,862	10,565	22%	15,985	13,413	19%
R&D expense	5,145	4,226	22%	8,268	7,258	14%
Net loss	1,654	2,021	–18%	3,767	4,134	–9%
Industry						
Number of companies	317	294	8%	1,274	1,287	–1%
Employees	94,000	73,000	29%	140,000	118,000	19%

Source: Ernst & Young, 1998a.

FIGURE 3.1 U.S. net trade balance: biotechnology, 1990-1996. Source: NSF, 1998.

4

LIKELY FUTURE POSITION

Assessment of the publication impact of immunologists and the results of the reputation survey clearly indicate that the United States is in a leadership position in the world in essentially all subfields of immunology. Given current trends, it seems likely that this position will be maintained for the next 5-10 years. However, several factors can adversely affect this position. There are four potential threats to US leadership in immunology:

- Funding and resource limitations.
- Increased competition from Europe and other countries.
- Clinical immunology and the shift toward HMOs.
- Training of US students.

Each potential threat is discussed below.

4.1 Funding and Resource Limitations

Current optimism as to the sustained US leadership in immunology is based in large part on a positive attitude toward NIH in the US Congress. That attitude is indicated by the proposals in the last year to double the NIH budget in the next 5-10 years. It must be recognized, however, that this could change. A return to the funding situation of the late 1980s and early 1990s, with low pay grades and administrative cuts in funded-grant applications, could possibly harm the US leadership position by driving investigators and students away from biomedical research in general. It must be recognized that, despite important contributions from the biotechnology and pharmaceutical industries, NIH remains the engine that drives immunology.

The current practice of protecting intellectual property has the potential to restrict the two-way flow of information between academic institutions and the biotechnology industry in the life sciences, including immunology. That applies to reduction in sharing both research materials and information. If the situation occurs on a broad scale, opportunities to explore promising research projects might be restricted.

The growth in the number of material transfer agreements (MTA) that are often overlegalistic and protective of the broadest possible outcomes of the use of potentially proprietary materials has spawned technology-office bureaucracies in industry and in academic institutions; these offices can delay material transfer for months. It would be of great use if a simple, direct, legally binding, universal MTA for both industry and academe could be created and ratified by agreement or use.

The increasing cost of maintaining mouse facilities has raised serious concern among academic researchers. Although the cost of the mice is reasonable, as is the cost of the component of their care that includes husbandry, housing, feeding, and cleaning, as long as the charges match the costs on a species-specific basis, very large increases in charges often result for the following reasons: specialized veterinary care, which for all species is usually distributed in a species-nonspecific fashion, as are administrative and staff costs; the increased personnel efforts that are required to meet regulatory-compliance needs; and Office of Management and Budget (OMB) indirect-cost allotments.

For example, one US biomedical institution switched from non-species-specific allocation of costs to species-specific allocations (using an independent accounting firm) and lowered mouse charges by 30-40% (Stanford Medical School, 1998). Its former assessed charges exceeded by a factor of 2-5 the actual costs at institutions that use only mice for their research. The high mouse charges are common in the United States, but most laboratories in Europe and Japan are costed more directly or are subsidized. If this trend continues, many US researchers will have great difficulty in financially supporting mouse facilities.

Actions by major funding agencies could relieve much of the burden: First, all costs and resulting charges could be strictly species-specific. Second, cost-accounting for simple husbandry could be separated from that for veterinary-intensive care. Third, efforts to simplify (and, when appropriate, eliminate) regulatory-compliance requirements could be undertaken. Fourth, the A-21 set of guidelines from OMB regarding indirect cost charges for federally-funded research could be reevaluated as to whether animal facilities can be removed from the special-services category, so that indirect costs could be lowered.

4.2 Increased Competition from Europe and Other Countries

In many countries, there appears to be a trend away from the customary hierarchical systems of funding, research, and employment of scientists toward the US system of competitive peer review. There also appears to be a trend toward better funding from government and private agencies and an increasing emergence of the biotechnology industry in many European countries. Together, those factors will enhance the quality of non-US immunology and make it more competitive.

4.3 Clinical Immunology and the Shift Toward HMOs

The clinical impact of immunology has long been limited by clinical subspecialization. For example, although the clinical practice of allergy is separate from other aspects of clinical immunology (such as rheumatology), basic and clinical research in the two fields overlap extensively. Until recently, clinical immunology barely existed as a definable field. Although the situation had shown signs of improving, reports (May et al. 1997; Campbell et al. 1997) indicate that the increasing dominance of HMOs in funding medical care in the United States potentially has an increasingly adverse effect on clinical research in general and clinical immunology in particular. This are several reasons. For example, HMOs compete for patients with academic clinicians, and this means that fewer patients are available for academic clinical trials; this poses a loss of a source of income that has traditionally been a source of funding for academic clinical research and a concurrent loss of jobs and opportunities for training of clinical immunologists.

Figure 4.1 shows the number of US citizens and permanent resident PhD students in immunology, and Table 4.1 and Figure 4.2 show the degree to which they are supported by NIH. As shown in Figure 4.1, the number of PhD students in immunology research has roughly doubled over the last 20 years. The percentage of these students supported by NIH has varied between 30 and 40% according to Table 4.1 and Figure 4.2.[2] Foreign students are not eligible to receive NIH training grants. The panel believes that this level of funding combined with the increasing time to degree and low wages influenced the quality of US students who entered immunology programs.

4.4 Training of US Students

Panel members perceive the quality of US graduate students and postdoctoral fellows in immunology to be declining. Several factors

[2] Data in this paragraph from special analysis by NRC Office of Scientific and Engineering Personnel of data from the survey of Doctorate Recipients and the Survey of Earned Doctorates for this study.

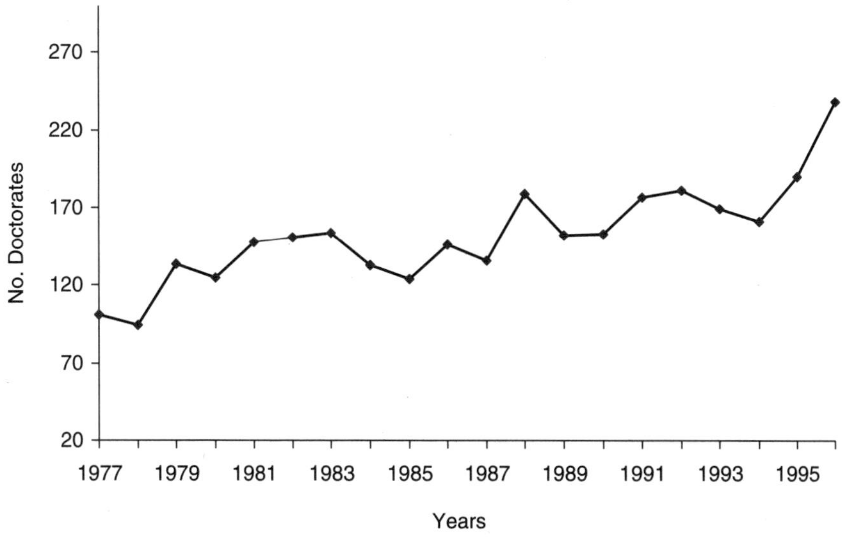

FIGURE 4.1 Number of PhD students in immunology in the United States, 1977-1996. Source: Analysis conducted by National Research Council's Office of Scientific and Engineering Personnel of Survey of Doctorate Recipients for this study.

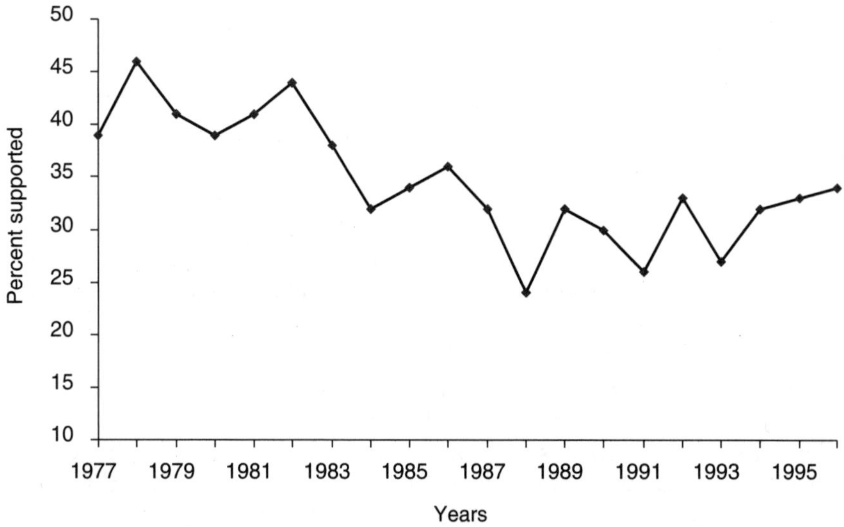

FIGURE 4.2 Percentage of US citizen and permanent-resident PhD students in immunology supported by National Institutes of Health, 1977-1996. Source: Analysis conducted by National Research Council's Office of Scientific and Engineering Personnel of Survey of Earned Doctorates for this study.

TABLE 4.1 NIH Trainee and Fellowship Support in Immunology

Year	No. of Doctorates	No. of Citizens or Permanent-Resident Doctorates	No. of Doctorates Supported by NIH	No. of Citizens or Permanent-Resident Supported by NIH	Fraction of Citizen or Permanent-Resident Doctorates Supported By NIH
1977	101	90	39	39	43%
1978	94	86	43	43	50%
1979	134	131	56	55	42%
1980	125	119	49	49	41%
1981	148	141	61	60	43%
1982	151	136	66	66	49%
1983	154	137	58	58	42%
1984	133	121	43	42	35%
1985	124	113	42	42	37%
1986	146	129	54	53	41%
1987	136	113	45	44	39%
1988	179	164	43	43	26%
1989	152	136	49	49	36%
1990	153	129	46	46	36%
1991	177	140	47	46	33%
1992	181	155	60	60	39%
1993	169	131	47	46	35%
1994	161	143	51	51	36%
1995	190	171	62	62	36%
1996	238	198	81	80	40%

Source: Analysis conducted by National Research Council's Office of Scientific and Engineering Personnel of Survey of Earned Doctorates for this study.

might contribute to a decline in quality. The trend toward department structures in which students are admitted into a large multidisciplinary program before choosing a specialty offers more varied opportunities for students. Because immunology is often, although inaccurately, viewed as too specialized and less interdisciplinary than other fields, students might be choosing other fields that are considered more general. Graduate study (of 5-7 years) followed by 3-5 years of postdoctoral training at salaries less than those of technicians might lead many talented young US citizens to choose other fields of endeavor. There is also a loss of MD talent in the field because of the cost of education and the salary differentials after completion of degree work.

In the United States, while there has been a downward trend in the number of PhD immunologists in academic positions, there has been a steady increase in the number of non-tenure-track appointments as shown in Table 4.2. In the early 1980s, 50% of immunologists with

TABLE 4.2 Employment Status of Doctorates in Immunology

	1973	1975	1977	1979	1981	1983	1985	1987	1989	1991	1993	1995
Tenured and Tenure Track Faculty	0	6	22	146	146	195	166	209	181	306	458	423
Tenured Faculty	0	0	0	1	1	40	101	118	100	137	181	180
Tenured Track Faculty	0	6	22	145	145	155	65	91	81	169	277	243
Other Academic Position	4	9	21	44	105	167	154	260	385	357	430	445
Postdoc Appointments—Academic	0	10	92	131	40	141	172	173	276	305	251	211
2 Year College Faculty	0	0	24	0	0	12	10	2	0	0	0	18
Industry	0	22	0	15	48	82	238	255	316	260	419	556
Federal and Other Government Positions	0	0	11	0	39	24	29	56	97	91	128	161
Self Employed and Others	0	7	48	58	106	126	63	97	93	293	97	311
Postdoc Appointments—Other	9	2	11	20	15	59	114	34	89	89	152	155
Unemployed and Seeking	0	5	0	0	34	1	2	0	12	0	42	10
Elementary and High School Teachers	0	0	0	0	0	0	0	0	0	0	30	18
TOTAL	13	61	229	414	533	807	948	1086	1449	1701	2007	2308
Tenured and Tenure Track Faculty	0.0%	9.8%	9.6%	35.3%	27.4%	24.2%	17.5%	19.2%	12.5%	18.0%	22.8%	18.3%
Tenured Faculty	0.0%	0.0%	0.0%	0.2%	0.2%	5.0%	10.7%	10.9%	6.9%	8.1%	9.0%	7.8%
Tenured Track Faculty	0.0%	9.8%	9.6%	35.0%	27.2%	19.2%	6.9%	8.4%	5.6%	9.9%	13.8%	10.5%
Other Academic Position	30.8%	14.8%	9.2%	10.6%	19.7%	20.7%	16.2%	23.9%	26.6%	21.0%	21.4%	19.3%
Postdoc Appointments—Academic	0.0%	16.4%	40.2%	31.6%	7.5%	17.5%	18.1%	15.9%	19.0%	17.9%	12.5%	9.1%
2 Year College Faculty	0.0%	0.0%	10.5%	0.0%	0.0%	1.5%	1.1%	0.2%	0.0%	0.0%	0.0%	0.8%
Industry	0.0%	36.1%	0.0%	3.6%	9.0%	10.2%	25.1%	23.5%	21.8%	15.3%	20.9%	24.1%
Federal and Other Government Positions	0.0%	0.0%	4.8%	0.0%	7.3%	3.0%	3.1%	5.2%	6.7%	5.3%	6.4%	7.0%
Self Employed and Others	0.0%	11.5%	21.0%	14.0%	19.9%	15.6%	6.6%	8.9%	6.4%	17.2%	4.8%	13.5%
Postdoc Appointments—Other	69.2%	3.3%	4.8%	4.8%	2.8%	7.3%	12.0%	3.1%	6.1%	5.2%	7.6%	6.7%
Unemployed and Seeking	0.0%	8.2%	0.0%	0.0%	6.4%	0.1%	0.2%	0.0%	0.8%	0.0%	2.1%	0.4%
Elementary and High School Teachers	0.0%	0.0%	0.0%	0.0%	0.0%	0.0%	0.0%	0.0%	0.0%	0.0%	1.5%	0.8%

Source: Analysis conducted by the National Research Council's Office of Scientic and Engineering Personnel of Survey of Doctorate Recipients for this study.
Note that this is a sample study.

academic appointments had tenure or were in a tenure-track. In 1995, the proportioned had decreased to about 40%. In the last 15 years, there has been an even more rapid increase in immunologists in industrial careers. Only about 10% of PhD immunologists went into industrial positions after completing their training in 1981, and almost 25% in 1995. The unemployment rate has remained very low[3]. Data for comparisons with other countries were unavailable.

[3] Data in this paragraph from special analysis by NRC Office of Scientific and Engineering Personnel of data from the Survey of Doctorate Recipients for this study.

5

SUMMARY AND CONCLUSIONS

The multiple assessment methods used in this study resulted in very similar findings, although the benchmarking process was limited by a scarcity of rigorous and unbiased data. A scientifically rigorous benchmarking process was not possible, because independent and unbiased adequate data could not be identified. The methods, despite their flaws, were the best available to us. Their results, with the panel's judgment, support the conclusions presented below.

5.1 The United States Is the World Leader in All the Major Subfields of Immunology but Is Only Among the World Leaders in Some Specific Sub-Subfields.

On the basis of the results of three benchmarking methods—a virtual-congress survey, citation analysis, and publication counts—the United States appears to be preeminent in immunology. Furthermore, it leads in all four of the subfields examined: cellular immunology, molecular immunology, immunogenetics, and clinical aspects of immunology. That is not a surprising result, given the size of the US enterprise, which writes some 60% of the immunology papers published each year. However, what is of greater interest, given the size of the enterprise, is that in some sub-subfields the United States is only among the world leaders: lymphocyte development and self-nonself recognition, inherited immunodeficiency, tumor immunology, and transplantation and immunosuppressive drugs.

The inherent flaws of each method make rigorous and exact assessment of US leadership impossible. However, these approaches do yield an estimation of the present standing of the United States in immunology. Because the exploration of immunology is an international endeavor, the high degree of cooperation and collaboration among US and non-US scientists should be highlighted.

Current US leadership has been documented by a number of quantitative and semiquantitative measures, but these measures do not show the breakthrough discoveries that are recognized by such awards as the Nobel Prize (half the immunology awardees were non-citizens). Nor do they reveal that a very significant fraction of the leading US scientists received some of or all their training in non-US institutions, mainly as nationals in other countries; several of these were Nobel laureates for research done outside of the United States.

5.2 Flexibility To Pursue Original and Innovative Research Ideas Has Attracted Both Domestic and International Human Capital. Federal, State, and Private Funding Have All Contributed to a Climate Ripe for This Innovative Research.

The United States has been able to attract talented foreign students to be both graduate and postgraduate investigators in immunology laboratories to a greater degree than other countries have been able to attract US students. That is in part due to the research opportunities available within the United States for these students as they seek to advance their careers. In the United States, more than in other countries, high-school and college students have the opportunity to gain research and analytical experience by working in laboratories and attending specialized science programs.

The NIH has been the major federal funding agency for immunology research. The strength of this system is that it is largely an investigator-initiated, peer-reviewed, and merit-based system of awarding grants. Critically, it is the individual investigator—rather than the department chair or other research colleagues, as it often is in many European countries—that has the authority and autonomy to pursue a specific research interest. Unlike many foreign countries, the United States supports research institutions and medical schools through state governments and private foundations, and this allows the freedom and flexibility to develop innovative research programs.

5.3 Industrial Interests Have Fostered Many Striking Breakthroughs in Immunology.

Substantial funding of the biotechnology industry by venture capitalists and other investors has resulted in the successful generation of many products to sell in the international market. Venture-capital financing of the biotechnology industry increased by 11.7% from $697 million in 1996 to $790 million in 1997. (BIO, 1997; BIO, 1998) In addition to creating an economic benefit to the United States, the success of the US biotech industry has resulted in the creation of new jobs for immunology graduates. And, the collaboration between academic and industrial researchers has allowed scientific discoveries to be rapidly developed and commercialized, in contrast with what has been observed in many other countries.

5.4 A Scarcity of Large-Scale Clinical Trials in Immunology Can Be Attributed to Shortages of Funding and of Qualified Personnel. In Addition, Increasing Dominance of Managed Care Means That Fewer Patients Are Available to Academic Institutions for Clinical Trials.

The expense of a large-scale clinical trial often proves prohibitive, especially when there is fierce competition among institutions and between research interests for limited funding dollars. European countries, because of their centralized government control of medical schools and research institutions have been able to support large-scale clinical trials more successfully than the United States. Anecdotal evidence indicates a decrease in trained clinical immunologists to serve as principal investigators for such trials in the United States. Lack of funding and training opportunities has contributed to the growing scarcity. Furthermore, the advent of the managed care has decreased the patient base for this type of clinical research.

5.5 Shifting Federal and Industry Priorities, a Potential Reduction in Access to Domestic and Foreign Talent, and the Increasing Cost of Maintaining Mouse Facilities Could Curtail US Ability To Capitalize on Leadership Opportunities.

Continued US leadership in the various subfields of immunology is not guaranteed. It depends on trends and sudden changes in the United States and abroad in funding, human resources, and infrastructure support. NIH has received increases in its annual budget from Congress, and the increases have resulted in the funding of more investigator-initiated grants in many fields of research, including immunology.

The trend of creating multidisciplinary graduate programs at large universities has resulted in competition for immunology graduate students and postdoctoral fellows. In addition, there is a substantial decrease in medical doctors seeking to specialize in immunology, in part probably, because of the cost of such an education and the low salary offered during the training period. Other countries, particularly those in Europe, seem to be moving away from the restrictive funding and tight employment environments that have been characteristic of their scientific research institutions. That raises the possibility that foreign students will elect to seek training and jobs in their own respective countries. The loss of talented students in immunology, both domestic and international, would have profound implications for the ability of the United States to maintain its leadership role.

One subject of particular concern to the panel was the lack of adequate funding and specific cost-based accounting for maintaining mouse facilities at most research institutions. Because much immunology research involves the use of mice, this resource is critical to the development of the field.

6

REFERENCES

Bio, 1998. *The 1998-99 Bio Editors and Reporters Guide to Biotechnology. http://www.bio.org*

Bio, 1997. *The 1997-98. Bio Editors and Reporters Guide to Biotechnology. http://www.bio.org*

Ernst & Young. *European Life Sciences 98 : Continental Shift,* London, United Kingdom: Ernst & Young. April 1998b.

Ernst & Young. *Biotech 1999: Bridging the Gap.* Palo Alto, California: Ernst & Young. December 1998a, *http://www.ey.com.*

May, Ernest, Anthony Mazzashi, Rebecca J. Levin, David Blake, and Paul Griner. 1997. *Relationship Between National Institutes of Health Research Awards to US Medical Schools and Managed Care Presentation.* Journal of the American Medical Association, 278: 3.

NSF (National Science Foundation). *Science and Engineering Indicators*. NSB 98-1. Washington D.C.: 1998.

Stanford Medical School, 1998. Conversation Between Irving Weissman and Michael Hindery, Associate Dean, Stanford Medical School regarding Cost of Animal Care Facilities.

Weissman, Joel S., David Blumenthal, and Eric Campbell. 1997. *Relationship Between Market Competition and the Activities and Attitudes of Medical School Faculty.* Journal of the American Medical Association. 278:3.

APPENDIX

PANEL AND STAFF BIOGRAPHICAL INFORMATION

Panel

Irving L. Weissman *(Chair)* received his MD from Stanford University in 1965. He pursued training in experimental pathology at Oxford University and continued his postgraduate fellowship at Stanford University. Dr. Weissman is the Karel and Avice Beekhuis Professor of Cancer Biology, professor of pathology, professor of developmental biology, and, by courtesy professor of biological sciences at Stanford University. He has received the Outstanding Investigator Award from the National Institutes of Health, the Pasarow Award for Outstanding Contribution to Cancer Biology and the Harvey Lecture, and the Montana Conservationist of the Year Award. He is a member of the National Academy of Sciences, the American Association for the Advancement of Science, the American Academy of Arts and Sciences, the California Academy of Medicine, and the Israel Immunological Society. He has served as president of the American Association of Immunologists. He was the cofounder of the biotechnology companies Systemix, Inc., and Stem Cells, Inc.

James Allison received his PhD from the University of Texas, Austin in 1973. He did postdoctoral work at the Scripps Clinic and Research Foundation in La Jolla. Dr. Allison is professor of immunology, director of the Cancer Research Laboratory, and a Howard Hughes Medical Investigator at the University of California, Berkeley. Selected awards and honors include a merit award from the National Institutes of Health and, election into the National Academy of Sciences and the American

Academy of Microbiology. He serves as a councilor to the American Association of Immunologists and is a member of the Board of Scientific Counselors for the National Cancer Institute.

Frederick W. Alt received his PhD in biological sciences at Stanford University, where he worked with Robert Schimke and discovered the phenomenon of gene amplification in the context of cellular resistance to anticancer drugs. In 1982, he joined the faculty of the College of Physicians and Surgeons of Columbia University in New York, where he became professor of biochemistry and molecular biophysics and, professor of microbiology. In 1987, he became a Howard Hughes Medical Investigator at Columbia University. In 1991, Dr. Alt became senior investigator at the Center for Blood Research in Boston, in addition to serving as a Howard Hughes Investigator at Boston's Children's Hospital. He is professor of genetics and pediatrics at Harvard Medical School, chair of the NIH allergy and immunology study section and of the Irvington Institute Scientific Advisory Board. He is a member of the National Academy of Sciences, the American Academy of Microbiology, and the American Academy of Arts and Sciences. Among his many honors are the Irma T. Hirschl Career Scientist Award, the Searle Scholars Award, the Mallinckrodt Scholar Award, and an NIH Merit Award.

Harald von Boehmer studied medicine at the Universities of Gottingen, Frieburg and Munich and prepared his medical thesis at the Max Planck Institute for Biochemistry. He is an adjunct professor in the Department of Pathology, University of Florida, Gainsville, and professor of immunology, University of Basel, and, Faculte de Medecine Necker Enfants Malades, Descartes University, Paris. He is the director of 373 of the National Unite Institute of Science and Medical Research, France. He is a member of the Institut Universitaire de France, Academia Europaea, the European Molecular Biology Organization, the New York Academy of Sciences, Gesellschaft fur Immunologie, the American Association of Immunologists, and the Scandinavian Society for Immunology. Dr. von Boehmer has been awarded the Louis Jeantet Prize for Medicine, the Avery-Landsteiner Prize for Immunology, the Paul Ehrlich and Ludwig Darmstaedt Prize, and the Korber Prize for European Science. He chairs the Executive Committee of the European Journal of Immunology.

Max D. Cooper received his MD (1957) and training in Pediatrics (1958-1960) at Tulane Medical School. He was a house officer and research assistant at the Hospital for Sick Children, London(1960-1961), and a pediatric-allergy fellow at the University of California, San Francisco Medical Center (1961-1962). His postdoctoral research in the laboratory of Robert Good (1963-1967), led to the definition of separate T- and B-cell lineages. Dr. Cooper is professor of medicine, pediatrics, pathology, and microbiology at the University of Alabama at Birmingham; senior scientist at the University of Alabama Comprehensive Cancer Center; professor of medicine and director of the Division of Developmental and Clinical Immunology at the University of Alabama; and a Howard Hughes Medical Investigator. He is a member of the National Academy of Sciences and in 1990 was elected to the Institute of Medicine. He was inducted as a fellow in the American Association for the Advancement of Science. Dr. Cooper served as president of the American Association of Immunologists and of the Clinical Immunology Society. Among his awards are the 3M Life Sciences Award, the Sandoz Prize for Immunology, and the American College of Physicians Award.

Irwin Feller is the director of the Institute for Policy Research and Evaluation and professor of economics at the Pennsylvania State University, where he has been on the faculty since 1963. Dr. Feller was an American Society for Mechanical Engineering Pennsylvania State Fellow for 1996-1997. Dr. Feller's research interests include the economics of academic research, the university's role in technology-based economic development, and the evaluation of federal and state technology programs. He was chair of the Committee on Science, Engineering, and Public Policy, American Association for the Advancement of Science.

Laurie H. Glimcher received her MD at Harvard Medical School in 1976. She was an intern and resident at Massachusetts General Hospital and a postdoctoral fellow under the direction of William Paul at the National Institutes of Health. Dr. Glimcher is a physician in the Division of Rheumatology and Immunology at Brigham and Women's Hospital, professor of medicine at Harvard Medical School, and Irene Heinz Given Professor of Immunology at the Harvard School of Public Health. She received a Merit Award from NIH, was elected into the American Academy of Arts and Sciences, and received the Lee S. Howley Award from the Arthritis Foundation. She serves on the corporate board of directors for Bristol-Myers Squibb. She is a councilor of the American Association of Immunologists.

David V. Goeddel received his PhD in biochemistry in 1977 from the University of Colorado in Boulder. He was a postdoctoral fellow at the Stanford Research Institute. Dr. Goeddel is the president and chief executive officer of Tularik, Inc. He is a fellow of the American Association for the Advancement of Science and a member of the American Academy of Arts and Sciences, the National Academy of Sciences, and the American Academy of Microbiology. Dr. Goeddel serves on the editorial review boards of Immunity and Nature Biotechnology. His research interests include cytokine signaling mechanisms and small-molecule therapeutics that act through regulation of gene expression.

Hugh McDevitt received his MD from Harvard Medical School in 1955. He was an intern in medicine at Peter Bent Brigham Hospital, a resident in medicine at Bellevue Hospital, and a postdoctoral fellow in the Department of Bacteriology and Immunology at Harvard Medical School. Dr. McDevitt is professor of medicine and of microbiology and immunology at Stanford University School of Medicine. He has received the 3M Life Sciences Award, the Paul Erlich Prize, and Outstanding Investigator Award from NCI and NIH, the Barbara Davis Diabetes Award, and the Paul Klemperer Award from the New York Academy of Sciences. He became a member of the National Academy of Sciences in 1977, of the Institute of Medicine in 1983, and of the Royal Society of London in 1994.

Diane Mathis received a doctorate in biology from the University of Rochester, New York in 1977. She is the directeur of research, INSERM, LGME, and Institut de Genetique et de Biologie Moleculaire et Cellulaire (IGBMC) in Strasbourg, France. She serves on the editorial boards of the *European Journal of Immunology*, *Immunology Today*, *Comptes Rendus de l'Academie des Sciences de Paris*, *Science*, *Cell*, *Current Biology*, *Journal of Experimental Medicine*, and *Immunity*.

Gustav Nossal studied medicine at the University of Sydney and after 2 years of residency at the Royal Prince Alfred Hospital moved to Melbourne to work as a research fellow at the Walter and Eliza Hall Institute of Medical Research, where he received a PhD. Apart from 2 years as an assistant professor of genetics at Stanford University, 1 year at the Pasteur Institute in Paris, and 1 year as a special consultant to the World Health Organization, Sir Nossal's research career has been at the Hall Institute. He was the director of the institute from 1965 until he retired in 1996. Sir Nossal was also professor of Medical Biology at the University of Melbourne. Sir Nossal's eminence in immunology has been recognized by his election as president of the 25,000-member International Union of Immunological Societies. Included among his international honors is his election to the US National Academy of Sciences and his membership in the Academie des Sciences (France). He has also served as president of the Australian Academy of Science and chair of the global programme for vaccines and immunization of the World Health Organization. Sir Nossal was knighted in 1977 and made a Companion of the Order of Australia in 1989.

Roger M. Perlmutter received his MD and PhD from Washington University (St. Louis) in 1979. Thereafter, he pursued clinical training in internal medicine at the Massachusetts General Hospital and the University of California, San Francisco. He was a lecturer in the Division of Biology at the California Institute of Technology, where he studied the genetic basis of antibody repertoire diversification. He joined the departments of medicine and biochemistry and the Howard Hughes Medical Institute at the University of Washington (Seattle), where he became professor and founding chair of the Department of Immunology. In 1997, he left the University of Washington to assume responsibility for drug-discovery efforts at the Merck Research Laboratories in Rahway, NJ. Dr. Perlmutter has served on numerous scientific advisory and review panels and is a councilor of the American Association of Immunologists and a member of the Board of Directors of the Federation of American Societies for Experimental Biology.

Craig B. Thompson received his MD from the University of Pennsylvania in 1977. His internship and residency were at the Peter Bent Brigham Hospital. Dr. Thompson is a professor in the Department of Medicine and Molecular Genetics and Cell Biology at the University of Chicago and a Howard Hughes Medical Investigator. He has received the Jerome W. Conn Award for Distinguished Research by a Junior Faculty Member. He serves on the editorial boards of *Cell*, *Immunity*, and *International Immunology*.

Don C. Wiley was an NSF graduate fellow in biophysics at Harvard University and received his PhD in biophysics in 1971. Dr. Wiley is a professor of biochemistry and biophysics at Harvard University, a Howard Hughes Medical Investigator, a research associate in medicine at the Boston Children's Hospital, and an affiliate of the Department of Chemistry and Chemical Biology at Harvard University. He has been elected to numerous honorary societies, including the American Academy of Arts and Sciences, the American Association for the Advancement of Science, and the National Academy of Sciences. Among his awards are the Louisa Gross Horwitz Prize, the William B. Coley Award for Distinguished Research in Fundamental Immunology, the V.D. Mattia Award, the Passano Foundation Award, the Emil von Behring Prize, the Gairdner Foundation International Award, the Albert Lasker Basic Medical Research Award, and the Rose Payne Distinguished Scientist Award.

Staff

Deborah D. Stine is the study director and associate director of the Committee on Science, Engineering, and Public Policy (COSEPUP). She has worked on various projects throughout the National Academy of Sciences complex since 1989. She received a National Research Council group award for her first study for COSEPUP on policy implications of greenhouse warming and a Commission on Life Sciences staff citation for her work in risk assessment and management. Other studies have addressed graduate education, responsible conduct of research, careers in science and engineering, environmental remediation, the national biological survey, and corporate environmental stewardship. Dr. Stine received a PhD in public administration, specializing in policy analysis, from the American University. Before coming to the Academy, she was a mathematician for the US Air Force, an air-pollution engineer for the state of Texas, and an air-issues manager for the Chemical Manufacturers Association.

Tamara Zemlo is a Cancer Prevention Fellow at the National Cancer Institute (NCI), where she is researching the risk factors for the progression of low-grade cervical disease to cervical cancer. She is also participating in analyzing data from the ASCUS/LSIL Triage Study, which is an NCI-sponsored clinical trial designed to determine the optimal management plan for low-grade cervical cytologic abnormalities. She received a PhD in oncology from the University of Wisconsin—Madison, where she studied the transforming properties of papillomavirus replication proteins in tissue culture, and a Master's of Public Health from Harvard University. As part of her postdoctoral training, she has an internship at COSEPUP.